The Trade in Papers Marked with Non-Latin Characters / Le commerce des papiers à marques à caractères non-latins

Islamic Manuscripts and Books

VOLUME 15

The titles published in this series are listed at *brill.com/imb*

The Trade in Papers Marked with Non-Latin Characters / Le commerce des papiers à marques à caractères non-latins

Documents and History / Documents and Histoire

VOLUME 1

Edited by

Anne Regourd

TUTA SUB AEGIDE PALLAS · 1683 ·

BRILL

LEIDEN | BOSTON

Cover illustration / Illustration de couverture:
Ms. LMSG00005. Papier SAMIER BE-NAZEER au lion et soleil © Photographie Sara Fani, projet European Research Council (ERC) « Islam in the Horn of Africa. A Comparative Literary Approach », Advanced Grant no. 322849, Université de Copenhague.

Library of Congress Cataloging-in-Publication Data

Names: Regourd, Anne, editor.
Title: The trade in papers marked with non-Latin characters : documents and history. Volume 1 – Le commerce des papiers a marques a caracteres non-latins : documents et histoire. Volume 1 / edited by Anne Regourd.
Other titles: Commerce des papiers a marques a caracteres non-latins : documents et histoire. Volume 1
Description: Leiden ; Boston : Brill, [2018] | Series: Islamic manuscripts and books, ISSN 1877–9964 ; volume 15 | In English and French. | Includes bibliographical references and index.
Identifiers: LCCN 2017057552 (print) | LCCN 2018024015 (ebook) | ISBN 9789004360877 (E-book) | ISBN 9789004357402 | ISBN 9789004357402 (hardback :alk. paper) | ISBN 9789004360877 (e-book)
Subjects: LCSH: Paper industry—History. | Papermaking—History. | Watermarks. | Trade routes. | Manuscripts.
Classification: LCC TS1090 (ebook) | LCC TS1090 .T725 2018 (print) | DDC 676—dc23
LC record available at https://lccn.loc.gov/2017057552

Typeface for the Latin, Greek, and Cyrillic scripts: "Brill". See and download: brill.com/brill-typeface.

ISSN 1877-9964
ISBN 978-90-04-35740-2 (hardback)
ISBN 978-90-04-36087-7 (e-book)

This book is printed on acid-free paper and produced in a sustainable manner.

Contents

Indices

Foreword

At the end of the first millennium paper blew over as a new writing material from the Middle East to Africa, and from there the art of papermaking spread to Europe. Paper mills and paper production were established in the 11th century by Arab papermakers in Xàtiva on the Iberian Peninsula. Paper production spread to Italy in the 13th century, where papermakers in Fabriano adapted the technology and production of rag paper to a new rag paper quality, supplied with a watermark as the logo of the paper mill. The new mechanical treatment of rags by water-driven stampers, inspired by the treatment of woollen fabrics in fulling mills, instead of pounding by hand, resulted in a stronger and more durable writing material, because of the fibrillation of the fibre material and the internal sizing with gelatine that replaced the Arabian sizing of the paper surface with starch paste.

The importance of the watermark as a unique mark for the origin and date of paper has been studied and registered by filigranologists since the end of the 19th century. The present publication *The Trade in Papers Marked with Non-Latin Characters* illustrates by the registration and study of watermarks that European and especially Italian handmade and machine-made paper were imported in Islamic intellectual centres of manuscript cultures in the 17th–20th century. The watermarks in laid and wove paper as well as the impressed watermarks in machine-made paper with Non-Latin characters document that numerous Italian, Russian and other European paper mills produced writing paper especially for these customers in the Ottoman markets from the Red Sea, Yemen, Ethiopia as far as Nigeria in West Africa, as well as in the Mediterranean, Bilād al-Šām, and Iran.

When considering the many paper mills at the end of the first millennium along the Eastern and Southern Mediterranean regions, the articles presented here add a new chapter to the history of paper. The authors illustrate through their studies of the Islamic manuscripts and registration of the watermarks that the demand for durable and strong paper qualities, fit for writing with ink and pen, was met by the imported watermarked European paper. The Paper Road circled back from Europe to the Islamic centres of manuscript production in Africa and the Middle East, because paper sized with gelatine was better suited to the climate than the original Arabian paper, sized with starch.

Anna-Grethe Rischel
Paper Conservator emeritus, National Museum of Denmark
President of International Association of Paper Historians (IPH)

Remerciements

Je remercie les éditions Brill et les directeurs de la collection « Islamic Manuscripts and Books », Arnoud Vrolijk et Christoph Rauch, d'avoir accueilli ce livre à la fois historique et technique.

Toute ma gratitude va aux contributeurs, qui ont accordé leur confiance au projet dès sa naissance sans jamais s'en départir sur la longueur. Mme Anna-Grethe Rischel a aimablement accepté d'en rédiger la préface.

Je suis reconnaissante au projet ERC « Islam in the Horn of Africa: A Compartative Literary Approach », Université de Copenhague, de son soutien à mon travail sur ce livre.

Enfin, j'adresse une pensée particulière aux amis et aux collègues qui, au cours du temps, ont porté l'entreprise, chacun à sa manière.

List of Illustrations

Maps

1 http://www.lib.utexas.edu/maps/historical/africa_1885.jpg (accessed 4 August 2015).

Notes on Contributors

Michaelle Biddle
is paper and book conservator at Wesleyan University Library and is consulting conservator for northern Nigerian manuscript collections. She has published on Nigerian manuscript inks, and preservation efforts in *The Journal of Islamic Manuscripts* and has several forthcoming book chapters on various watermarks and the paper trade.

Evyn Kropf
MSc (2009), University of Michigan, is Librarian for Near Eastern & Religious Studies and Curator of the Islamic Manuscripts Collection at the University of Michigan Library. She has published several contributions on the material aspects of Islamic manuscripts, including "A Conservative Tradition? Arab Papers of the 12th–17th Centuries from the Islamic Manuscripts Collection at the University of Michigan" in the *Journal of Islamic Manuscripts*.

Anne Regourd
Ph.D. (1987), is Senior Researcher at the University of Copenhagen, ERC "Islam in the Horn of Africa", and an Associate at the CNRS, France. She has published extensively in the fields of History and Philology dealing with Codicology, Paper studies, and Papyrology.

Francis Richard
is member of the Research Unit UMR 7528 "Mondes iranien et indien" (Paris), former curator at the French National Library, former Head of the Department of Arts of Islam at The Louvre, and former Scientific Director of BULAC Library. He published extensively on Iranian manuscripts.

Alice Shafi-Leblanc
M.D. in History of Islamic Arts (2011), Paris IV-La Sorbonne, is a free-lance researcher. She wrote two dissertations on Muzaffarid (Iran, Fars, 1353–1393) manuscripts: *Recherche sur les manuscrits illustrés mozaffarides* (2010) and *La production de manuscrits enluminés mozaffarides entre 1355 et 1409: aspects codicologiques et ornementaux* (2011). Her current contribution is her first publication.

Jan Just Witkam
(Leiden, 1945) is professor emeritus of the University of Leiden, The Netherlands, with the chair 'Manuscript culture of the Islamic World'. He has widely published on Islamic book culture. He is editor-in-chief of *Journal of Islamic Manuscripts* (Brill, Leiden).

Olga Yastrebova
Ph.D. (2011), St. Petersburg State University, is a researcher and conservator of Oriental manuscripts at the National Library of Russia, and part-time docent at the Oriental Department of St. Petersburg State University.

Transliteration System for Arabic

The transliteration system for Arabic follows widely the system of *Arabica*[1] or ISO 233–232: 1993[2] and the Bibliothèque nationale de France[3] with the following modifications:

1. Consonants

أ	ʾ	ض	*ḍ*
ب	*b*	ط	*ṭ*
ت	*t*	ظ	*ẓ*
ث	*ṯ*	ع	ʿ
ج	*ǧ*	غ	*ġ*
ح	*ḥ*	ف	*f*
خ	*ḫ*	ق	*q*
د	*d*	ك	*k*
ذ	*ḏ*	ل	*l*
ر	*r*	م	*m*
ز	*z*	ن	*n*
س	*s*	ه	*h*
ش	*š*	و	*w*
ص	*ṣ*	ي	*y*

No word-initial *hamza.*

2. Vowels
a/*ā*, *i*/*ī*, *u*/*ū*.

3. Diphthongs
ay, *aw*.

1 http://www.brill.com/sites/default/files/ftp/authors_instructions/ARAB.pdf.
2 https://www.iso.org/obp/ui/fr/#iso:std:4118:fr.
3 http://guideducatalogueur.bnf.fr/ABN/GPC.nsf/C00F8804C7C3E372C12576A8002BED96/$FILE/EXTTranslitteration%20arabe.htm#_Toc248306778.

4. Definite Article
No distinction is made between solar and lunar letters (e.g.: *al-šams, al-qamar, Kitāb iḥyā' 'ulūm al-dīn*).

5. *tā' marbūṭa*
Final *tā' marbūṭa* is transliterated *-a* (e.g.: *risāla*).

6. Possessive or genitive constructions (*iḍāfa*)
In possessive or genitive constructions, the *tā' marbūṭa* followed by a complement is transliterated *-at* (e.g.: *Risālat al-tawḥīd*).

7. Prepositions preceding definite and indefinite nouns or pronouns
Whenever prepositions are prefixed to definite and indefinite nouns or to pronouns, they are separated by a hyphen. The article (*al-*) is maintained in full (e.g.: *fī al-bayt, bi-al-'aql, li-al-mamlūk, bi-allatī, li-allaḏīna*).

8. Conjunctions *wa-* and *fa-*
Hyphens are also used between such conjunctions as *wa-* and *fa-* and definite and indefinite nouns. The article (*al-*) is maintained in full (e.g.: *wa-al-mušāhada, fa-al-mušāhada*).

9. Prepositions and conjunctions before word-initial *hamza*
Such nouns are maintained in full (e.g.: *li-istiqbāl, wa-istiqbāl, li-amr, wa-amr*).

10. Suffix Pronouns
Suffix pronouns are not separated from nouns (i.e. no separation marks) (e.g.: *kitābuhu, kitābuhā*).

11. *Šadda*
Double *yā'* in relative adjectives and abstract nouns is transliterated as follows: *'arabī*, but *'arabiyya, 'inda al-Faransiyyīn, al-insāniyya.*

12. Personal names
Articles are fully maintained. *Ibn* is abbreviated *b.* and followed by the genitive; however, is kept in full, *Ibn*, etc. *Ibn* whenever it comes first in personal names (e.g.: Abū al-Barakāt, Ibn al-Hayṯam, 'Alī b. Abī Ṭālib).
At the start of a sentence or paragraph the "a" of the article "*al-*" is capitalised.

Introduction: Le papier des manuscrits, une source pour l'histoire du commerce

Anne Regourd

Dès 1985, Terence Walz ouvrait la voie à l'étude du commerce du papier à partir de papiers filigranés relevés dans des documents d'archives égyptiens pour un ouvrage collectif sur le Soudan. Ce travail remarquable n'a pas immédiatement fait école. Dans « Les routes commerciales entre Zabîd et l'Europe : les papiers filigranés de fonds manuscrits de Zabîd (Yémen, fin 18e–milieu 20e s.) », en 2006, nous avons repris le projet de reconstitution des routes de trois papiers identifiés dans des manuscrits de la ville de Zabid, puis, en 2011–2012, effectué une première tentative, suivie d'autres, pour ériger l'étude de Walz, inaugurale mais régionale, en méthode[1].

Le papier comme source

La méthode ainsi définie implique un saut épistémologique conséquent qui consiste à considérer les papiers des manuscrits comme source primaire d'une histoire des échanges du papier. Dans un premier volume de la série « Documents et Histoire », nous attirions l'attention sur l'importance pour la papyrologie des données physiques – aussi bien que textuelles – dans l'analyse des documents[2]. Si l'on en croit l'historiographie française, depuis le xviie s., l'École des Annales et d'autres efforts en faveur d'une ouverture de la définition de ce qui fait source historique – ou de ce qui a valeur documentaire, l'attachement des historiens à l'écrit demeure, semble-t-il, grand. Le caractère documentaire de ce qui n'est pas de l'écrit a pourtant été à nouveau souligné par les historiens des années 1920 qui ont théorisé la notion de document et était auparavant attesté par la création de corpus qui pouvaient ne pas être

1 Voir A. Regourd, « Les routes commerciales entre Zabîd et l'Europe : les papiers filigranés de fonds manuscrits de Zabîd (Yémen, fin 18e–milieu 20e s.) », 2006. Puis, *id.*, « Paper Trade on the Red Sea (XIXe c.–first half of the XXe c.): 'local papers' against Italian's. The case of an 'Ottoman' watermarked paper used in Yemen and in Ethiopia ».

2 A. Regourd, Introduction dans : *Documents et histoire. Islam, viie–xvie siècle*, 2013.

 | DOI 10.1163/9789004360877_002

composés prioritairement d'écrits – nous pensons par exemple à l'histoire orale ou aux données archéologiques[3]. Plus récemment, l'étude des textiles et vêtements a évolué dans ce sens, si bien qu'il est difficile de voir aujourd'hui leur apport documentaire réservé à une histoire du textile et du vêtement ou bien d'envisager l'étude d'un objet épigraphié sans analyser son support[4]. La codicologie, science auxiliaire de l'histoire, prend en considération les éléments matériels des codex pour reconstituer une histoire du livre : l'identification des papiers, le temps écoulé entre leur production et leur utilisation, d'autres aspects encore font déjà partie de son champ. Les informations délivrées par les papiers servent aussi à écrire une histoire des techniques. Ici, le statut épistémologique des papiers fait d'eux plus qu'un élément de critique externe[5] des sources, il en fait l'une des données de l'histoire du commerce, de la migration des hommes, des idées et des religions.

Méthodologie

L'étude technique comparée des papiers des manuscrits, croisée avec d'autres données, aide à l'identification de la fabrique d'origine ou à l'accumulation des informations qui rendent certains lieux de fabrication ou de départ improbables, ou mieux, exclus. Suivre cette méthode exige la constitution d'un corpus par la sélection soigneuse de manuscrits, avec, comme premiers critères, l'assurance d'une date et – une information moins fréquemment apparente – d'un lieu de copie. C'est alors que le papier peut être classé comme « papier à destination ». Une trace de son passage est saisie. Au-delà de cas uniques ou isolés, la réalisation de nuages statistiques pour l'utilisation d'un papier donné confirme la présence d'un marché durant une période. La fourchette de datation une fois consolidée, applicable aux manuscrits non datés écrits sur le même papier, accroît la base de l'analyse d'autant.

D'autres données sont fournies par la seule étude des papiers présents dans les manuscrits, elles donnent des indices sur l'identification des papiers exportés et leurs destinations, l'équipement des fabricants, leur capacité de production et d'ajustement au marché, la grandeur des feuilles à la livraison, voire leur

3 Voir l'œuvre fondatrice de J. Carcopino, *La vie quotidienne à Rome à l'apogée de l'Empire*, 1939.

4 Voir l'initial et exemplaire E. Kühnel & L. Bellinger, *Catalogue of Dated Tiraz Fabrics: Umayyad, Abbasid, Fatimid*, 1952.

5 N. Offenstadt *et alii*, *Historiographies : Concepts et débats*, 2010, 70*sq*.

conditionnement, enfin la compétition qui les met aux prises les uns avec les autres. Le coût et le prix du papier, les quantités importées et les montants, les modes et voies d'acheminement – sans chercher à être exhaustif – impliquent d'autres types de sources, qui renseignent aussi sur l'histoire des fabricants et de leurs moulins, les réseaux et les marchands impliqués dans le commerce du papier. Dans cette quête, les papiers des manuscrits représentent une source prometteuse qu'il convient certes de croiser avec d'autres sources, mais dont l'apport et la valeur euristique ont été insuffisamment exploités.

L'étude paraissait faisable dans des zones climatiques variées. Le papier est un bien de consommation paradoxal : à la fois périssable et consommable, de même que la nourriture ou les vêtements, il est aussi, précieux et conservé avec soin, un bien de consommation durable. Les manuscrits de l'Afrique subsaharienne en sont un bon exemple. Il s'agit d'une denrée rare jusqu'au xixe s., on le réutilise ; il abrite des textes précieux à protéger ; différents moyens de conservation des codex manuscrits existaient avant l'ère contemporaine, bref, un certain nombre de facteurs l'ont aidé à traverser les siècles.

Papiers à caractères non-latins

Les papiers à caractères non-latins paraissaient un bon point de départ comme possibles témoins, soit de lieu de production, avec à la clé la possibilité d'une redistribution des cartes par la multiplication des lieux de fabrication hors d'Europe, soit de destination, en direction d'un marché donné. Pour le monde arabe, se pose la question d'une « sortie » de la production papetière italienne, prédominante à partir du xviie s. Mais l'enquête supposait alors de travailler sur une période où la Révolution industrielle battait son plein, d'aborder des papiers produits à la machine, peu étudiés, du moins dès que l'on touche aux manuscrits islamiques. En revanche, c'est à l'époque moderne et contemporaine que les manuscrits sont les plus nombreux : ce sont ceux qui sont parvenus jusqu'à nous et, pour certaines régions, telle l'Afrique, il a été constaté que la production manuscrite (aussi bien codex que comptabilité de marchands) a connu un essor au xixe s., consécutif à la présence accrue de papier[6]. Au Yémen, si l'on extrapole à partir des collections de Zabid, Taez et Shibam

6 Voir Gh. Lydon, *On Trans-Saharan Trails. Islamic Law, Trade Networks, and Cross-Cultural Exchange in Nineteenth-Century Western Africa*, 2009, 102*sq.*, pour la Mauritanie. Au même moment, le même phénomène est observé en Occident.

(Hadramaout) et de quelques catalogues, le gros de la production manuscrite aujourd'hui conservée date de la seconde moitié du xviie au xxe s.

La découverte du papier à marque en alphabet arabe (ou arabo-persan) « al-Ǧ/ǧalālī » trouvé dans un manuscrit daté du xvie s. montre que le phénomène commence bien plus tôt qu'on ne le pensait (Shafi-Leblanc). Le plus ancien exemple connu en alphabet arabe, arabo-persan ou turc-ottoman, était jusqu'alors le filigrane composé d'un croissant et d'une étoile à huit branches accompagné de la contremarque *Islâmbôl sene 1219*, soit 1804–1805, publié par Ş. Tekin[7]. Les autres papiers étudiés dans ce volume sont bien du xixe et du xxe s., le plus récent est présent dans un manuscrit de 1386/1967. Certains des papiers décrits ont été utilisés pour des impressions typographiques de livres – concurremment à la fabrication de codex et de documents manuscrits, comme feuillets de garde ou font office de papier à en-tête.

Deux exemples de vernaculaires utilisés dans les échanges commerciaux pour désigner des papiers, l'un, *Alikurna*, attesté dès le xviie s., l'autre, Abū Šubbāk, en 1886, repris ensuite comme contremarque en écriture et langue arabe (avec peut-être des caractéristiques physiques appropriées), par des producteurs italiens, respectivement au xixe s. et au xxe s., lient la production d'un papier à une aire de consommation donnée et confirment l'existence de papiers produits exclusivement pour l'exportation (Kropf, Regourd 1, Witkam) : un vernaculaire naît chez les consommateurs d'un papier importé de qualité, très apprécié, il devient ensuite une appellation ou un générique, toujours dans la langue du marché, puis il renaît dans les marques du papier d'un fabricant dont ce n'est pas forcément la langue. Une préférence pour tel ou tel papier est en tous les cas exprimée, en première ligne par scribes et copistes, elle se traduit par une élasticité à la consommation réduite. De nombreux fabricants ou marchands considèrent visiblement, quant à eux, qu'utiliser la langue des consommateurs africains et du monde arabe (ottoman ?) est un bon argument de vente. On affectionnait les papiers colorés, en particulier en bleu et en vert, au Moyen-Orient et en Iran, un goût que la production russe n'a pas manqué de noter, ainsi que l'illustre le cas des papiers à timbres secs marqués en cyrillique, exportés vers l'Iran entre 1850 et 1880 (Richard).

---

7 Ş. Tekin, *Eski türklerde yazı, kâğıt, kitap ve kâğıt damgaları*, 1993, 90–91. Autres exemples dans V. Nikolaev, *Watermarks of the Mediaeval Ottoman Documents in Bulgarian Libraries, Watermarks of the Ottoman Empire*, 1954, vol. 1, n° 938, années 1807/1808 ; 944, 1808 ; 945, 1808 ; 946, 1808 ; 947, 1808 ; 977, 1812.

Un sujet d'histoire globale

Le commerce du papier est typiquement un sujet d'histoire globale. Jusqu'à la Révolution industrielle, les centres de production de papiers exportés en direction des mondes arabes et africains sont occidentaux avec une nette prédominance italienne dès le xviie s. La mécanisation du secteur conduit-elle à un *new deal*, à l'émergence d'autres centres de production et à une concurrence accrue ? L'Inde de l'Est et le Népal ont produit des papiers à pâte végétale (Thymelaceae sp.) ainsi que des papiers à base de chiffon dans l'Ouest, probablement dès le xiie s., et les papiers à matériaux de récupération mixtes (chiffons, fibres et papier usagés) étaient en forte proportion jusqu'en 1875 ; la présence de moulins à papier actionnés mécaniquement est attestée dès le premier quart du xviiie s. et connaît une accélération au xixe s. certes[8]. Mais les marchands de Bombay font de l'import-export de papiers anglais avec leur « logo » et trafiquent sur l'océan Indien dans les rets de la Compagnie des Indes au tournant du xxe s. (Regourd 2). En Turquie, la production de papier est instable (Kropf, Regourd 1), il faut attendre les années 1933–1938 pour qu'elle soit incluse dans un plan quinquennal[9]. En Iran, les papiers industriels à timbres secs portant caractères cyrilliques, exportés de Russie en Iran dans la deuxième moitié du xixe s. et utilisés dans l'ensemble du pays, supplantent insensiblement les papiers locaux (Richard).

Pour l'Afrique, nous avons des exemples de marchands et affréteurs, certains de familles juive nord-africaine et levantine, basés à Londres, Manchester et Liverpool, qui, bénéficiant de connections et d'établissements au Moyen-Orient et en Afrique de l'Ouest, parfois en Occident, faisaient entre autre de l'exportation de papier vers le Nigéria entre les deux dernières décennies du xixe s. et le premier tiers du xxe s. (Biddle 1). Quant à la péninsule Arabique, au Yémen, pays où la copie de manuscrit est importante, les papiers machine Abū Šubbāk cartonnent dans la première moitié du xxe s. ; l'un d'eux, produit par les célèbres moulins Galvani, parvient en Éthiopie. La production de papier n'échappe pas à la dynamique coloniale d'importation de matières premières : l'alfa/esparto d'Afrique du Nord est une plante entrant dans la composition de la pâte, fort appréciée des papetiers dès le milieu du xixe s., car elle produit un papier de qualité (Biddle 2). L'un des grands importateurs en est le Royaume-Uni.

---

8 S. Ramaseshan, « The History of Paper in India up to 1948 », 1989, 105*sq.*, 116*sq.*

9 M. A. Kâğıtçı, *Historical Study of Paper Industry in Turkey/Historique de l'industrie papetière en Turquie*, 1976.

Le leadership de la production de papier demeure occidental : d'après les firmes identifiées – mais toutes ne le sont pas, la compétition se joue entre des firmes de l'hémisphère Nord, au Moyen-Orient, en Afrique et dans la péninsule Arabique. Les papetiers n'ont pas tous pris la décision d'équiper leurs moulins au même moment. Le ms. IES 4505, conservé à Addis Abeba, en Éthiopie, daté au colophon de ǧumādā al-awwal 1237/janvier-février 1822 (f. 89v), contient 5 cahiers formés d'un papier sans marque à pâte à bois, mais aussi un quinion de papier italien EAN, pour un seul et même texte. Le papier anglais fait une apparition notable sur les « terres » des papiers italiens. L'importance des marchands d'import-export non-producteurs de papier, mais proches du « terrain », que signale l'apparition de leur marque dans le papier en Afrique et en Inde, marque qu'ils faisaient partager parfois à d'autres produits, est l'un des phénomènes les plus significatifs de la fin du xixe à la première moitié du xxe s. À moindre échelle, plus subrepticement, la commande de copie de manuscrits par des Occidentaux (Witkam) prend, dans un pays comme l'Éthiopie, une signification particulière avec l'entrée du papier dans églises et monastères.

Les routes du papier

Les marchés ciblés représentent les premières balises des routes du papier. Les papiers des manuscrits ou papiers « à destination » nous aident, eux, à tracer des voies de pénétration multiples (revendeurs, boutiquiers, caravanes ...), parfois informelles ou imprévisibles (individuelles), bref difficilement détectables dans les archives des fabricants, à partir de premiers grands points de livraison ; ils suggèrent des réseaux d'échange. C'est le cas du papier *Alikurna* au xixe s., utilisé dans une zone qui s'étend de la Turquie de l'Est à la Mésopotamie (Kropf). On retrouve des papiers anglais, commandés par des marchands indiens de Bombay, à Aden et au Harar, en Éthiopie : notre hypothèse est celle d'une redistribution depuis Aden en direction du Harar, plutôt qu'une livraison directe depuis Bombay[10]. Le réseau des commerçants Bohras pourrait alors en avoir été le fer de lance[11]. Reste à savoir si les papiers Abū Šubbāk d'Éthiopie

---

10 Ici, Regourd 2, et *id.*, « Papiers filigranés de manuscrits de Zabīd, premier tiers du XVIIIe–milieu du XXe siècle : papiers importés et 'locaux' », 2015, 250–251.

11 Sur les réseaux Bohras en Éthiopie et leurs liens au Yémen et à Bombay, voir D. Harre, « Exchanges and mobility in the Western Indian Ocean: Indians between Yemen and Ethiopia, 19th–20th centuries », 2017.

(types A et D) suivaient les mêmes étapes dans la première moitié du xxe s. Le Soudan semble avoir été livré directement. Ils demeurent un article typique de la mer Rouge.

Peu à peu surgit une carte des routes des papiers pour une période donnée, qui croise celle de la pénétration des religions du Livre. Elle est suivie de près par celle de leur utilisation, fort appréciable aux yeux du codicologue, lorsqu'il s'agit de localiser la production d'un manuscrit. Le papier est un marqueur spatial.

Papiers et pouvoir

Les papiers à marques en caractères cyrilliques sont connus, ils ont été décrits et catalogués. Mais le corpus formé par les *farmāns* du Shah iranien Fatḥ ʿAlī écrits sur du papier importé de Russie sur fonds de guerre russo-persane au Caucase (1804–1813) illustre les aspects subtils de l'économie de guerre (Yastrebova). Une réflexion similaire pourrait être faite au sujet du bureaucratique Empire ottoman, papivore, qui importait d'immenses quantités de papiers italiens. En 1860–1861, le ministère des Affaires étrangères tunisien utilise un papier filigrané – au lieu et place d'un papier à en-tête, servant à éviter les fraudes ; il est fait sur commande, probablement en France (Biddle 2). Le besoin en papier pour l'imprimerie, pour l'impression de livres, comme l'illustre le papier Abū Šubbāk B au Yémen du temps de l'Imam Yaḥyā b. Muhammad b. Yaḥyā Ḥamīd al-dīn (1869–1948, Regourd 1), mais surtout pour la presse, naissante, au xixe ou au xxe s. selon les pays, dans une ambiance de libéralisme politique toujours friande de pâte à bois, de nationalisme aussi, et, dans le cas yéménite, sous contrôle, est un autre sujet intéressant la question des importations d'état[12].

Ce livre offre une première contribution à la question du papier comme source pour l'écriture d'une histoire du commerce. Il reste encore beaucoup à faire dans l'ensemble, zones non-explorées et couvertes confondues.

12 Sur le contexte général et la question du libéralisme, de la presse et du pouvoir au xixe s. en Occident, voir F. Gaillard, « *Pulp story* ou Balzac médiologue », 1997, 212–213, « le souverain est partout, sauf sur le trône » (citation de Balzac, p. 212) ; « La liberté est directement proportionnelle à la consommation de papier » (p. 213).

Bibliographie

Carcopino, Jérôme. 1939. *La vie quotidienne à Rome à l'apogée de l'Empire*, Paris, Hachette, « La vie quotidienne ».

Fèvre, Lucien. 1992. *Combat pour l'histoire*, Paris, Armand Colin, « L'Ancien et le Nouveau ».

Gaillard, Françoise. Deuxième semestre 1997. « *Pulp story* ou Balzac médiologue », *Les Cahiers de médiologie* 4, « Pouvoirs du papier », 209–214.

Harre, Dominique. 2017. « Exchanges and mobility in the Western Indian Ocean: Indians between Yemen and Ethiopia, 19th–20th centuries », *Chroniques du manuscrit au Yémen*, numéro spécial 1, Anne Regourd & Nancy Um (éd.), « From Mountain to Mountain: Exchange between Yemen and Ethiopia, Medieval to Modern ».

Kâğıtçı, Mehmet Ali. 1976. *Historical Study of Paper Industry in Turkey/Historique de l'industrie papetière en Turquie*, Istanbul, Grafik Sanatlar Matbaası.

Kühnel, Ernst & Bellinger, Louisa. 1952. *Catalogue of Dated Tiraz Fabrics: Umayyad, Abbasid, Fatimid*, Washington.

Lydon, Ghislaine. 2009. *On Trans-Saharan Trails. Islamic Law, Trade Networks, and Cross-Cultural Exchange in Nineteenth-Century Western Africa*, Cambridge, Cambridge University Press.

Nikolaev, Veslovod. 1954. *Watermarks of the Mediaeval Ottoman Documents in Bulgarian Libraries, Watermarks of the Ottoman Empire*, Sofia, Bulgarian Academy of Sciences, vol. 1 [titre original en bulgare].

Offenstadt, Nicolas, *et alii*. 2010. *Historiographies : Concepts et débats*, Paris, Gallimard.

Ramaseshan, Sita. 1989. « The History of Paper in India up to 1948 », *Indian Journal of History of Science* 24/2, 103–121.

Regourd, Anne. 2006. « Les routes commerciales entre Zabîd et l'Europe : les papiers filigranés de fonds manuscrits de Zabîd (Yémen, fin 18e–milieu 20e s.) », dans : Renato Traini, *Convegno Storia e cultura dello Yemen in età islamica con particolare riferimento al periodo rasûlide (Roma, 30–31 octobre 2003)*, Rome, Bardi, 173–198.

Regourd, Anne. 2013. Introduction, dans : *Documents et histoire/Documents and History. Islam, viie–xvie siècle/Islam, 7th–16th century*, Actes des premières Journées d'étude internationales, École pratique des hautes études, IVe section-musée du Louvre, département des Arts de l'Islam, Paris, 16 et 17 mai 2008, Genève, Droz, « École pratique des hautes études, Sciences historiques et philologiques » II, 2013, 1–12.

Regourd, Anne. 2015. « Papiers filigranés de manuscrits de Zabīd, premier tiers du XVIIIe–milieu du XXe siècle : papiers importés et 'locaux' », Chap. 8 dans : David Hollenberg, Christoph Rauch & Sabine Schmidtke, *The Yemeni Manuscript Tradition*, Leyde/Boston, E. J. Brill, 227–251.

Regourd, Anne. À paraître. « Paper Trade on the Red Sea (XIXe c.–first half of the XXe c.): 'local papers' against Italian's. The case of an 'Ottoman' watermarked paper used in Yemen and in Ethiopia », *Festschrift in Honour of Professor Iraj Afshar*, Londres, Fondation al-Furqān.

Tekin, Şinasi. 1993. *Eski türklerde yazı, kâğıt, kitap ve kâğıt damgaları*, Istanbul, Eren Yayıncılık ve Kitapçılık.

Walz, Terence. 1985. « The Paper Trade of Egypt and the Sudan in the Eighteenth and Nineteenth Centuries », dans : Martin William Daly (éd.), *Modernization in the Sudan: Essays in Honor of Richard Hill*, New York, Lillian Barber Press, Inc., 29–48.

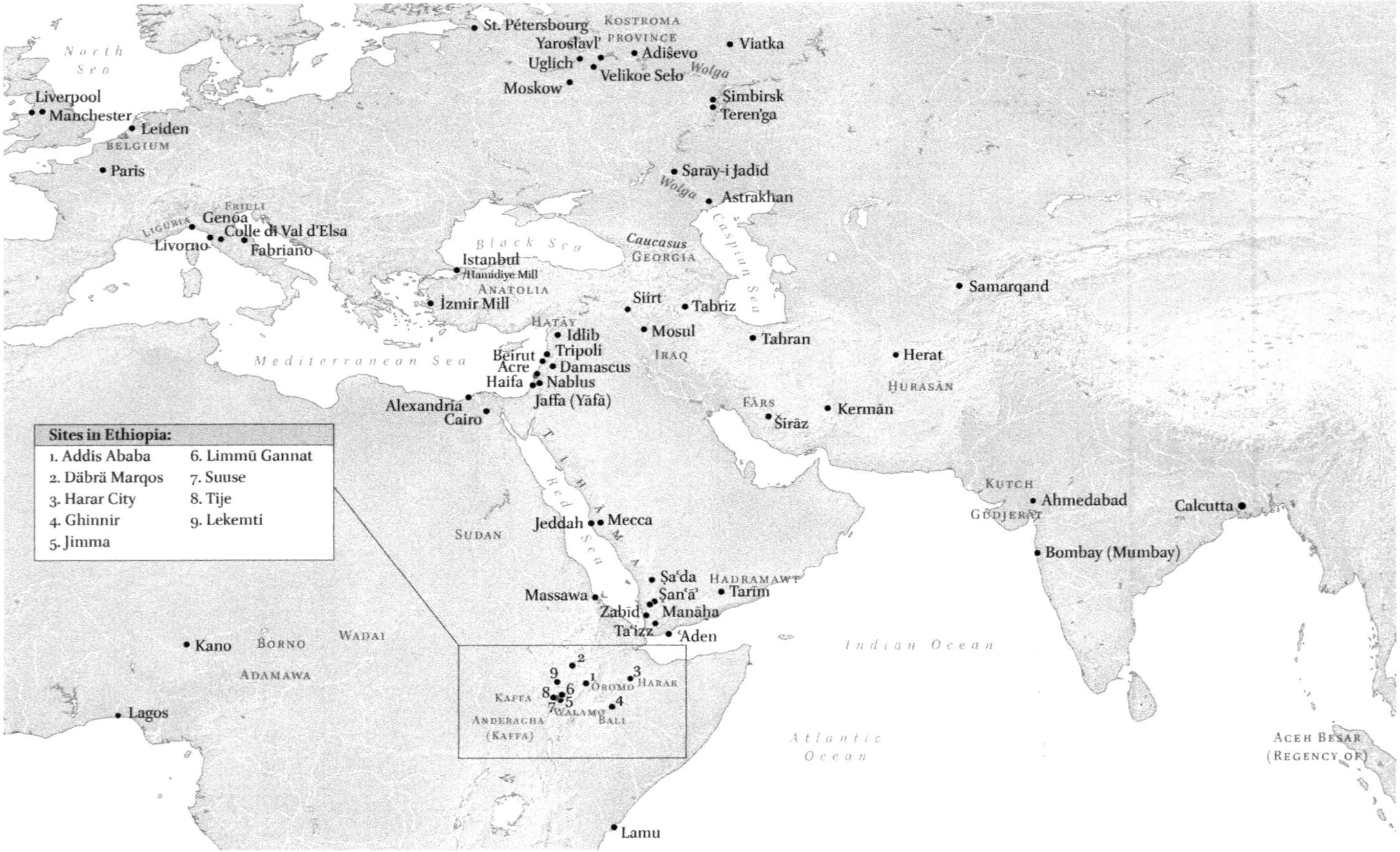

MAP 0.1 *Relevant places for paper production and trade referenced in this volume*

CHAPTER 1

ARBIB, YDLIBI and SÙRÙ (HAKURĪ): Three Arabic Script Watermarks in Northern Nigerian Manuscripts (end 19th–beginning 20th century)

Michaelle Biddle

Résumé

Trois filigranes portant un texte en écriture arabe ont été trouvés dans des manuscrits en écriture arabe, copiés au Nord du Nigéria. Il s'agit des filigranes :

1. *BENIAMINO ARBIB Yā Naṣīb*, « Yā Naṣīb », en arabe « Ô destinée, ô fortune ! », environ 1889–1919 ;
2. *A.YDLIBI 3 Brazil Street Manchester ʿAbd [a]l-Ġanī Idlibī*, « ʿAbd [a]l-Ġanī Idlibī » écrit en arabe, début du xxe s., 1921 ? ;
3. *SÙRÙ LÓGÙN AIYE*, en langue Yoruba et écriture arabe (*ʿağamī*), placé à l'intérieur d'une tablette à écrire, avec, de part et d'autre, *REGISTERED TRADE MARK*, habituellement lu : *Hakurī Māġanī Dūnyā* en langue hausa, écrite à l'aide de l'alphabet arabe (*ʿağamī*), 1908–1932.

Les trois marques apparaissent sur du papier fait à la machine. Cet article décrit comment elles ont été réalisées techniquement et discute les sources, dates et spécificités de chacun des papiers, ainsi que les changements de routes commerciales liés à l'arrière-fond politique des xixe et début du xxe s.

1 Introduction

Nigeria has a rich Islamic manuscript heritage that document the diverse intellectual and social history of West Africa and are of interest to scholars working in the fields of history, religion, Arabic grammar, poetry, astrology, mathematics, medicine, folklore, geography, education, logic, cultural studies, art history, archaeology, anthropology and jurisprudence. Whilst Muslim kingdoms were established in the Central Sudan, what is now called Northern Nigeria, by the eleventh century, surviving manuscripts date from centuries later. Few of these

 | DOI 10.1163/9789004360877_003

manuscripts are reliably dated. Identifying and dating the watermarks in the paper sheets provides clues as to when they were actually written, and identifying the mills that made the paper provide information about commercial trade and intellectual networks. These manuscripts are written in distinctive scripts variously called West African *Sūdānī* or *ʿIfriqī*, a heavy angular script that is an offshoot of North African *Maġribī*, a Western Kufic, and somewhat more cursive scripts called *ʿAǧamī* (also spelled *Ajami* or *Adjami*), often used for interlinear commentaries written in local languages of Hausa, Kanuri, Fulfulbe or Yoruba, and a similar script used by early nineteenth century jihadists, appropriately called *Ǧihādī*.[1]

There are major collections of these manuscripts in the Northern Nigerian cities of Sokoto, Kaduna, Kano, Jos, and others, and at Northwestern University in the United States, but Nigerian scholars believe that at least eighty percent still remain in private hands.[2] Three watermarks with Arabic script have been found in Arabic script manuscripts written in Northern Nigeria:

1. *BENIAMINO ARBIB Yā Naṣīb, Yā Naṣīb* in Arabic ('O fate, O fortune or O destiny')—ca. 1889–ca. 1919;
2. *A.YDLIBI 3 Brazil Street Manchester ʿAbd [a]l-Ġanī Idlibī*, "ʿAbd [a]l-Ġanī Idlibī" in Arabic script—early twentieth century, possibly 1921;
3. Yoruba phrase written in Arabic script (*ʿAǧamī*)—*SÙRÙ LÓGÙN AIYE* inside a wooden slate or writing board flanked by *REGISTERED TRADE MARK*, traditionally interpreted as Hausa *ʿAǧamī Hakurī Māġanī Dūnyā*—1908–1932.

2 Watermarking Machine-Made Paper with a Dandy Roll

All three of these watermarks appear on machine-made paper and as always the paper sheet is as informative as the watermark. In each case

1 A. D. H. Bivar, "The Arabic calligraphy of Nigeria", 1968, 3–15; J. O. Hunwick, *The writings of Central Sudanic Africa* (*Arabic Literature of Africa*. Vol. 2), 1995, 1–15; N. F. Safwat, *The art of the pen: calligraphy of the 14th to 20th centuries*, 1996, 229–230; id., *Golden pages: Qur'ans and other manuscripts from the collection of Ghason I. Shaker*, 2000, 284–287; S. Blair, *Islamic Calligraphy*, 2006, 571–575.

2 M. Biddle, *Saving Nigeria's Islamic manuscript heritage*, 2008, http://works.bepress.com/mbiddle/1/

these watermarks were made with a dandy roll attached to a continuous (Fourdrinier) machine.[3]

In the mid-1820s John Marshall of T. J. Marshall of London developed a cylindrical roller that could be attached to either a continuous or a cylinder machine. This cylindrical roller, or dandy roll, could make an impression of laid and chain lines upon the sheet of paper or pulp as it exited the machine. In 1830 Thomas Barrett of St. Mary Cray, London, obtained a patent for inserting the watermark and maker's name on continuous paper.[4] Dandy rolls were not in wide spread use until the 1840s. Initially watermarks were hand tied to the dandy roll, just like on a hand mould, but by the 1870s solder was used to attach the watermark to the wire cover of the roller and a common custom in England was that papers were "usually made on orders for a certain quality".[5] By the 1880s mills were advertising paper "specially watermarked to any design".[6] Other processes, such as Drury, Nash, and Brown, were developed to make watermark designs even more easily and cheaply. All these processes enabled individuals and firms to commission special designs watermarked on paper sheets.[7]

Because contact between the watermark on the dandy roll and wet sheet is usually fleeting, the impression can be less distinct and well defined than watermarks formed with a hand mould.[8] Dandy rolls and covers enable reams and reams of paper to be watermarked with the same image. However, because of uneven shrinkage and other factors, slight variations may be found from sheet to sheet. Dandy rolls, or their covers, can be easily changed and stored.

---

3 D. Hunter, *Papermaking: the history and technique of an ancient craft*, 1947, 400–408, covers much of the history, technique and issues in watermarking with dandy rolls. R. W. Sindall, *Paper Technology*, 1920, 28, 32–33, 271–273. Gacek identifies the Arbib mark as being an "impressed watermark (…) by a rubber stereo" (*Arabic Manuscripts: A Vademecum for Readers*, 2009, 130–131). This sort of marking leaves a ridge around the mark as it is a type of embossing but the Arbib mark is an integral part of the paper sheet, has no ridges and was made with a wire dandy roll on which the wires in the watermark was soldered to the wire cover of the dandy roll.

4 C. Tomlinson, ed. "Paper", 1855, 367.

5 Editor of "The Paper Mills Directory", *The art of papermaking: a guide to theory and practice*, 1874, 35.

6 *Papermills directory for England, Scotland & Ireland and year book of the paper-making trade for 1882*, Paper makers' announcements, 12.

7 A. P. Dillont, "Unusual developments in watermarking methods", 1916, 36; C. W. Hackelman, *Commercial engraving and printing: a manual of practical instruction and reference*, 1921, 639–641.

8 L. D. Smith, "The paper machine", 2003, http://www.motherbedford.com/watermarks/Watermark1B8.htm.

There is at least one documented case of one dandy roll being used intermittently over a ninety-three year period.[9]

3 Three Arabic Script Watermarks

1. *BENIAMINO ARBIB 'Yā Naṣīb'*, *Yā Naṣīb* in Arabic ('O fate, O fortune or O destiny')—ca. 1889–ca. 1919

FIGURE 1.1 BENIAMINO ARBIB, *Horseman brandishing a rifle*, "Yā Naṣīb" *in Arabic* MAF 182, MODIBBO FUFORE COLLECTION, AREWA HOUSE, AHMADU BELLO UNIVERSITY, KADUNA

Date:	ca. 1889–ca. 1919
Type:	laid machine-made
Watermark dimension:	72 mm diameter
Fiber:	hardwood, cotton
Condition:	brittle

9 C. L. Douglas, "New concepts in dandy roll design and construction for quality watermarking", 1970, 129. For more information on identifying machine-made paper and distinguishing between continuous and cylinder machine-made papers see M. Biddle, "New strategies in using watermarks to date sub-Saharan Islamic Manuscripts", 2017.

Color:	yellowish tan
Laid lines:	7 per cm
Chain lines:	24 mm
Bulk/thickness:	.11–.12 mm
Distance between watermarks:	128 mm

The Beniamino Arbib watermark is common in early twentieth century manuscripts written in Northern Nigeria. Northwestern University's 'Umar Falke Collection from Kano has 130 examples, amongst them Falke 14. Falke 14 is a Qur'an and also contains A. Ydlibi watermarked paper. Only Falke 793 is dated—1919, and it appears together with machine-made *Andrea Galvani Pordenone* paper with slender *Tre lune* ca. 1913.[10] The Arewa House Modibbo Fufore Collection from Adamawa has examples (MAF 37, 73, 182) and there also are many examples in the Arabic Manuscripts Collection at Jos Museum with those bearing dates being Jos 640 dated 1339/1922 and Jos 1439 dated 1904. A third, Jos 329, is dated 1889, the earliest dated example found.

According to the *Encyclopaedia Judaica*, the 'Arbibs' are a Jewish family of North African origin with Italian connections. By the last quarter of the nineteenth century the Arbib family was principal financial backers for trans-Saharan caravans having participated in the lucrative Tunisian esparto grass boom of the 1860s. As a replacement for rags, esparto grass was becoming an increasing important component in paper fabrication. Members of the family firm were located in Manchester, Tunis, and Tripoli with a network of agents and subagents scattered across Sudanic Africa, including Northern Nigeria.[11]

Beniamino Arbib was born in Benghazi, North Africa, 1854 and died in Manchester, Lancashire, England, 1919. Beniamino's father, Angelo Arbib left Tripoli in 1860 and settled in Cairo where he founded the Casa Arbib, which traded in ostrich feathers, ivory, rubber and leather. In the late 1870s Casa Arbib was still sending annual caravans from Tripoli to Wadai. Angelo died in 1891 and Beniamino took over the trading house.[12]

E. & E. Arbib are listed as shippers with a London address in 1870 but by 1877 they had addresses in London and Manchester with one of their addresses being 7 Brazennose Street, Manchester. In 1882 they are listed as Mediterranean, Alexandria and South of Europe shippers in colonial produce,

10 See M. Biddle, "New strategies in using watermarks to date sub-Saharan Islamic Manuscripts", 2017, where this particular mark has been labeled AGmm5.

11 R. Shenton, "A note on the origins of European commerce in Northern Nigeria", 2011, 169–170.

12 L. A. Balboni, *Gl'Italiani nella civilta egiziana del secolo XIX*, 1906, vol. 2, 106; G. Dalla Vedova, "Pellegrino Matteucci ed il sul diario inedito", September 1885, 543.

cottons, woolens, linens, furniture, tin, ammunition, and guns. In 1888 they are recorded as shipping esparto, a fiber of increasing importance to the paper trade, from Tripoli and Homos to London. Between 1889 and 1896 the Arbibs operated the SS Arbib Brothers cargo steam ship.[13]

At some point Beniamino took over the 7 Brazennose Street offices for his partnership with Clemente Hassan as 'shippers' at that address was dissolved 31 October 1892. Beniamino continued trading in Egypt as evidenced by his involvement with the importers board of trade in Alexandria. Beniamino was listed in the *1911 England Census* as resident in Manchester and his 1919 death is recorded in the *England and Wales Death Index 1916–2006*.[14]

At this time Manchester was the financial hub of Lancashire and the center of the cotton cloth manufacturing for Britain. Lancashire had the greatest concentration of paper mills in England due to its access to good coal and chemical resources, abundant supplies of reasonably clean, soft water and industrial labour. Lancashire also had a robust rail system with easy access to Liverpool, a major international seaport, trading extensively with Lagos, Nigeria.[15]

The paper watermarked with Beniamino Arbib's name is primarily of hardwood pulp.[16] Aspen, a hardwood, was imported into Britain at least as early as 1884 and the British Wood Pulp Associations was founded in 1897.[17] Hardwood fibers are well suited for producing smooth papers for printing and for writing and would be attractive to those writing with a reed pen. A more usual combination for English made paper of this period would be cotton, esparto

13 *Street's Indian and colonial mercantile directory for 1870*, 411; *Export merchant shippers of London, Manchester, Liverpool, Birmingham, Wolverhampton and Walsall, with their respective trading ports, and the class of goods they customarily ship*, 1877, 2; *The export merchant shippers of London, Manchester, Liverpool, Birmingham, Wolverhampton, Walsall, Leeds, Bradford, Sheffield, Glasgow, Greenock, Edinburgh, Leith, Hull, Newcastle-on-Tyne, Stockton-on-Tees, Middlesborough, West Hartlepool, Gateshead, Sunderland, North and South Shields, Bristol and Cardiff with their respective trading ports and the class of goods they customarily ship*, 1882, 77; *World's paper trade review*, 1888, 41, 104; *Marine Engineer, a monthly journal*, 1888–1889, 394; WRECKsite http://www.wrecksite.eu/wreck.aspx?155797.

14 *London Gazette* 4 November 1892, 6192, http://www.london-gazette.co.uk/issues/26341/pages/6192/page.pdf; B. Arbib, *Libro giallo relativo alle elezioni del Collegio del Commercio d'Importazione de 7 e 9 marzo 1904 in Alessandria-Egitto*, 1904.

15 A. D. Spicer, *The Paper Trade*, 1907, 189, 200–207; A. H. Shorter, *Paper making in the British Isles*, 1971, 146–160.

16 My thanks to Northwestern University Library Conservator Scott Devine and Joe Swider of McCrone Research Group, Chicago, for carrying out polarized light microscopy (plm) on the Arbib and Hakuri papers.

17 *World's paper trade review*, 1884, 376; *World's paper trade review*, 1897, 2.

and soft woods.[18] Beniamino Arbib was not manufacturing paper but merely taking advantage of modern machine paper production methods to have his name, as a merchant, impressed with a dandy roll on the machine-made paper he was selling. Paper branding was not new. It originated in the late thirteenth century with the development of the first watermarks in Italy.[19]

Walz posits that the *Yā Naṣīb* phrase (variously translated as 'O fate, O fortune or O destiny') refers to a Hausa lottery and speculates that the paper was created for it.[20] However in the Nigerian National Archives only the Enugu Archives of Eastern Nigeria branch has records of any public lotteries and they date from 1928 to 1966.[21] As Beniamino died in 1919 it would be remarkable if he was watermarking paper for a lottery that began nine years after his death. It could be that 'O fortune' is a reference to events of the period that impacted the Arbibs—the British take-over of Egypt in 1882, the collapse of the Mahdist government, with whom the Arbibs traded, or the impact of the vagaries of fate on the Arbib family. Walz also thinks that the horseman in the center of the watermark is brandishing a sword, but that would mean the horseman is holding the sword by the blade. It is more likely that the item is a rifle. In a 1938 paper Nabia Abbott misidentifies this watermark as seventeenth-eighteenth century, an impossibility as machine-made paper was not made until the nineteenth, but correctly identifies that the rider with flying robes as holding a gun. Also recent polarized light microscopy showing hardwood as the primary pulp contradicts Muṣṭafā al-Bābī al-Ḥalabī's 1976 statement in Cairo to Walz that the paper was made from rags.[22]

18 *Paper Mills Directory of England, Scotland and Ireland for 1871*, 3.

19 E. J. Labarre, *Dictionary and encyclopaedia of paper and paper-making*, 1952, 334.

20 T. Walz, "The paper trade of Egypt and the Sudan in the Eighteenth and Nineteenth Centuries", 1985, 42.

21 Nigerian National Archives Enugu Archives of Eastern Nigeria branch, http://archiveswiki.historians.org/index.php/National_Archives_of_Nigeria.

22 L. A. Balboni, *Gl'Italiani*, 1906, vol. 2, 106; T. Walz, "The paper trade of Egypt and the Sudan", 2011, 102–103, 103 note 75; N. Abbott, "Maghribī Koran manuscripts of the seventeenth to the eighteenth centuries", 1949, 61.

2. *A.YDLIBI 3 Brazil Street Manchester ʿAbd [a]l-Ġanī Idlibī*, "ʿAbd [a]l-Ġanī Idlibī" in Arabic script—early twentieth century, possibly 1921

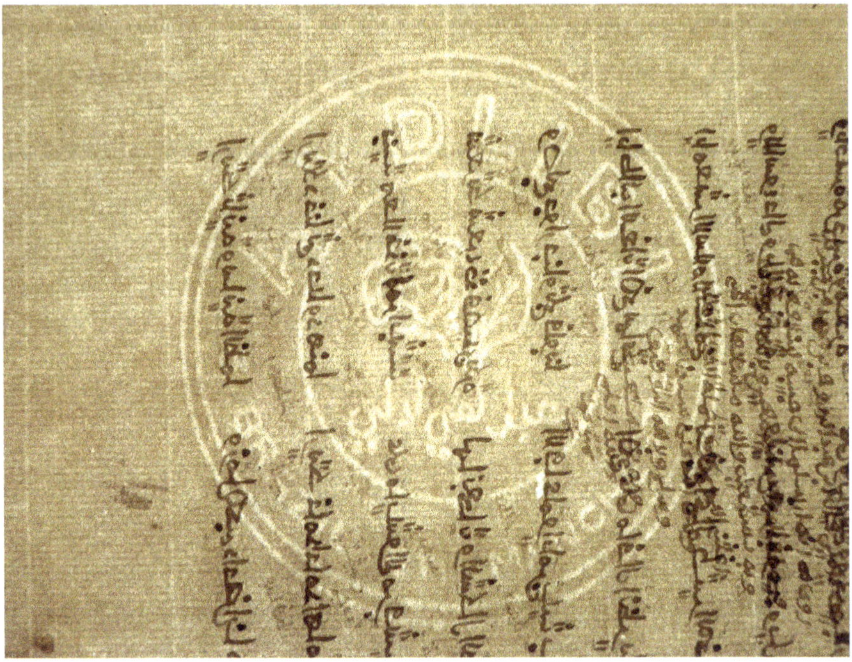

FIGURE 1.2 A. YDLIBI 3 BRAZIL ST. MANCHESTER *circling an Arabic inscription "ʿAbd [a]l-Ġanī Idlibī" underneath a flower, probably a rose*
FALKE 14, UMAR FALKE COLLECTION, MELVILLE J. HERSKOVITS LIBRARY OF AFRICAN STUDIES, NORTHWESTERN UNIVERSITY

Date:	ca. 1920s–30s, possibly ca. 1921
Type:	laid machine-made
Watermark dimension:	100 mm diameter
Fiber:	unknown
Condition:	brittle
Color:	dull tan
Laid lines:	8 per cm
Chain lines:	26.5 mm between chain lines
Bulk/thickness:	.11–.12 mm
Distance between watermarks:	unable to determine

This watermark has been found in Jos 654 and Falke 14, 82, 256, 438, 883, 2688. None of these manuscripts are dated. Falke 14 is a Qur'an and also contains Beniamino Arbib watermarked paper. Falke 1496 is dated 1354/1935.

Idlib (also transliterated Edlib or Idleb) is a city and province in Northwestern Syria, famous for an important cache of 4,500 year old cuneiform tablets found nearby at the ancient city of Ebla. The tablets were transferred to museums in Aleppo, Damascus and Idlib Museums.[23] Before the industry moved to nearby Aleppo, this city was also famous for its soap, often scented with rosewater.

Abdulgani Ydlibi (1887–1953) and Azaat Ydlibi were merchants and their surname suggests a Syrian origin. In 1917 they paid for some cotton goods carried by the Steamer: City of Birmingham. They are recorded as dissolving their Manchester, England, partnership with one Aref Namani on 26 January 1927. In 1930 they were involved in a court case Calico Printers v. Barclays Bank, where they are described as a merchant in Manchester with "an allied house or correspondent in Beyrout called Azaat Ydlibi".[24]

Abdulgani Ydlibi had a son with the same name born in 1921. This rare watermark could be a celebration of his son's birth, with the rose a sentimental or ancestral reference over the son's name spelled out in Arabic script. The Ydlibi business address of 3 Brazil Street, Manchester, is adjacent to Sackville Gardens, a city park with gardens and walks established in 1900. Abdulgani Ydlibi, the son, was an Royal Air Force (R.A.F.) pilot killed in action at Fuka, Egypt, in 1942. A personal memorial to the family is at Alderley Edge Cemetery in Cheshire, England, a suburb of Manchester.[25]

This merchant family had connections with Nigeria as in 1954 Abdulgani Ydlibi's younger son, Gordon, was a director of M. (Michael) & G. (Gordon) Ydlibi, Ltd., Exporters of Manchester, England, and made his home in Lagos. The company was dissolved in 1967.[26]

23 H. H. Wellisch, "Ebla: The World's Oldest Library", 1981, 488–500; G. Pettinato, *Ebla, a new look at history*, 1991.

24 *Steamer: City of Birmingham* 1917; *London Gazette* 1927, 688; *Calico Printers' Association*, 1930.

25 Find A Grave, http://www.findagrave.com/cgi-bin/fg.cgi?page=gr&GRid=52258940.

26 *Hartford Courant* (Connecticut) 22 September 1954, "Helen Zamulko, Gordon Ydlibi plan to marry", 12; *London Gazette* 1967, 13956.

3. Yoruba phrase written in Arabic script (*'Ağamī*)—*SÙRÙ LÓGÙN AIYE* inside a wooden slate or writing board flanked by *REGISTERED TRADE MARK*, traditionally interpreted as Hausa *'Ağamī Hakurī Māğanī Dūnyā*—1908–1932

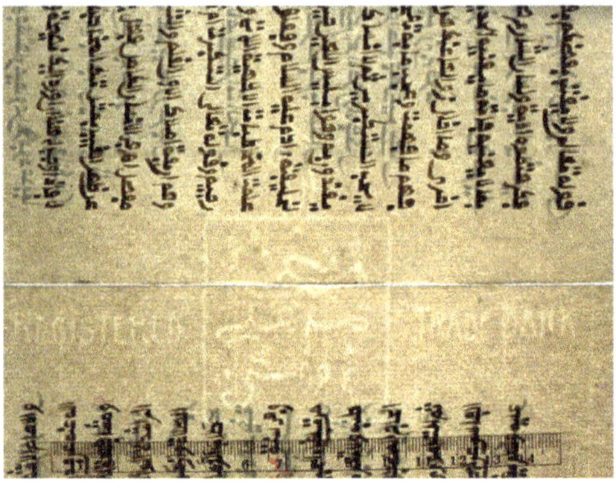

FIGURE 1.3 *Yoruba phrase written in Arabic script* ('Ağamī)—SÙRÙ LÓGÙN AIYE *inside a wooden slate or writing board flanked by* REGISTERED TRADE MARK, *traditionally interpreted as Hausa* 'Ağamī *"Hakurī Māğanī Dūnyā"*
FALKE 503, UMAR FALKE COLLECTION, MELVILLE J. HERSKOVITS LIBRARY OF AFRICAN STUDIES, NORTHWESTERN UNIVERSITY

Date:	1908–1932
Type:	laid machine-made
Watermark dimension:	84 mm high × 150 mm wide
Fiber:	grass, cereal straw fibers
Condition:	fair, smooth both sides
Color:	yellowish cream
Laid lines:	8 per cm
Chain Lines:	25–26.5 mm
Bulk/thickness:	.18–.22 mm
Distance between watermarks:	275 mm

This watermark is common in early twentieth century manuscripts written in Northern Nigeria. It also appears in Malian manuscripts.[27] The Arewa House Modibbo Fufore Collection from Adamawa has examples (MAF 191, 342).

27 For example, Ahmad Baba Institute, shelf mark 16089. The Ahmad Baba Institute collection is normally housed in Timbuktu, but due to the ongoing insurgency it is now in Bamako, Mali. I was at the ABI in April 2014 and saw four other manuscripts with this watermark but they were uncatalogued and had no shelf mark.

There are also examples in the Arabic Manuscripts Collection at Jos Museum (Jos 57, 214, 220, 304, 815, and 853). None of these Fufore or Jos manuscripts are dated. Northwestern University's Falke Collection, the intact library of the Kano trader ʿUmar Falke, who was also a local scholar and author, has numerous examples (Falke 140, 415, 503, 683, 685, 724, 788, 878, 1210, 1219, 1221, 1354, 1355, 1561, 1564, 1735, 1819, 2190, 2284, 2318, 2330, 2333 and others). None of these manuscripts are dated but many are in the hand of ʿUmar Falke, who died 22 Muḥarram 1382/25 June 1962.

The legend *REGISTERED TRADE MARK* flanks either side of a writing board or slate containing the Yoruba phrase *SÙRÙ LÓGÙN AIYE*, written in *ʿAǧamī*, i.e. an African language written in Arabic script. Since 1914 this phrase has been interpreted as a Hausa one written in *ʿAǧamī—Hakurī Māġanī Dūnyā* and been variously translated as "Patience is the remedy for everything"; "Forbearance is essential to happiness in this world" and "Patience is the way to cope with the world's ills".[28]

In actual fact this mark was registered as Trade Mark #305,306 on 7 August 1908 by the Lagos Stores, Limited, 6 Stanley Street, Liverpool. Lagos Stores self-identified as "West African Merchant and Shippers" and the translation they provided for this Yoruba phrase was "Patience conquers (the world) everything" [Fig. 1.4].[29]

The same day they also registered another paper mark, #305,305, "with the characters on Comb represent the Yoruba word *SÙRÙ*, meaning 'Patience'" [Fig. 1.5]. Whilst this paper mark has not yet been recorded in Northern Nigerian manuscripts in Arabic script, it might yet be found.[30]

United Kingdom trade mark registration lasts for ten years but both of these paper marks were renewed with no changes in 1922, fourteen years after the initial registration,[31] perhaps an indication of ambivalence on the profitability of continuing in the paper trade. The first appearance of this *SÙRÙ LÓGÙN AIYE* phrase inside a writing board and another somewhat similar one are contained within the labels of two 'Spirits (liquor)' trade marks [Figs. 1.6 and 1.7] originally registered in 1896 and renewed in 1910 when the company changed its name from The Lagos Stores and Tomlinson Ltd. to The Lagos Stores, Limited.

In 1897 Lagos Stores used the writing board with the Yoraba phrase *SÙRÙ LÓGÙN AIYE* [Fig. 1.4] as a trade mark for tobacco, food, porcelain, silk, mineral

28 C. H. Robinson, *Dictionary of the Hausa Language*. Vol. 1: *Hausa-English*, 1913, 134; G. P. Bargery, *A Hausa-English dictionary and English-Hausa vocabulary*, 1934, 439; T. Walz quoting Murray Last, "The paper trade", 1985, 48, note 54.

29 *Trade Marks Journal* 33, 1908, 1517.

30 *Trade Marks Journal* 33, 1908, 1867.

31 *Trade Marks Journal* 47, 1922, 1274.

FIGURE 1.4 *Trade Mark 305,306, 7 August 1908, for "Paper (except Paperhangings), Stationary, and Bookbinding"*
TRADE MARK JOURNAL 33, 1908, 1517

FIGURE 1.5 *Trade Mark 305,305, 7 August 1908, for "Paper (except Paperhangings), Stationary, and Bookbinding, but not including Stereotyping, Flong, Typewriters or Ribbons for Typewriters, and not including any Goods of a like kind to any of these excluded Goods"*
TRADE MARK JOURNAL 33, 1908, 1867

FIGURE 1.6 *A writing board or slate used in Trade Mark 192,896, 5 February 1896 for "Spirits"*
TRADE MARK JOURNAL 21, 1896, 381

FIGURE 1.7 *The same Yoruba phrase as in the 1908 papermark in a writing board used in Trade Mark 196,951, 29 July 1896 for "A Spirit"*
TRADE MARK JOURNAL 21, 1896, 938

water, and candles and soap. For all but the first of these, Lagos Stores specified that 'The characters on Mohammedan slate represent the Yoruba words SÙRÙ LÓGÙN AIYE, meaning 'Patience conquers (the world) everything'". Lagos Stores used this "Mohammedan slate" mark again in 1899, 1901 and 1903 for glass, linen, jute, clothing, arms and ammunition[32] and finally, in 1908, for the paper mark.

In spite of the decades long insistence of Lagos Stores that the phrase used in the watermark is Yoruba, the writing appears to be a case of exographia, writing which takes place in another language. According to Dmitry Bondarev, Head of West Africa Research Projects at the Centre for the Study of Manuscript Cultures at the University of Hamburg, "There are two peculiarities which make the interpretation a bit difficult, namely the *sukun* above the letter *rāʾ* at the end of the first line, and a dot at the end of the last line. Otherwise the transliteration is straightforward: '*ankur māganī dūniyya(n)*'. Given that *hakuri* has a dialect variant *ankuri*, the Hausa expression '*hakuri maganin duniya*' is a very close match to '*ankur māganī dūniyya(n)*', the literal translation being 'patience is a medicine of the world'. It is beyond my imagination how the Yoruba '*suru logun aiye*' could be rendered in writing as '*ankur māganī dūniyya(n)*' or how *suru* could be written as '*ankur*'. The most plausible explanation would be a translation of the Yoruba expression into a close Hausa equivalent just because the expression had to be in a written form. A similar practice can be found in Senegal where Fulfulde (Pular) is used as the only medium of writing when the speakers of other languages of Casamance need to write in Ajami."

Lagos Stores were general merchants exporting a wide variety of British made goods—porcelains, yarns, cottons, silks, mineral waters, spirits, arms and ammunition—to Nigeria and buying native produce for export—palm oil, ground nuts, cotton lint and hides. Lagos Stores had a branch in Kano as well as in Lagos by 1911.[33] By 1923 they had registered one hundred seventy-six trade marks. Other images used were a tortoise, bird, hedgehog, a *ġorfa* (Saharan granary) and a scene of an African family pounding yams. In the same year they began to combine these marks over the image of a tortoise and *THE LAGOS STORES, LIMITED* in a floral frame.[34]

A majority interest in The Lagos Stores, Limited was acquired in 1919, along with several other UK based West African and Nigerian trading firms by The African Association, Limited, when the latter assumed the name of The African

32 *Trade Marks Journal* 40, 1915, 1144–1146.

33 M. A. Modibbo, "The colonial state and European trading companies in Northern Nigeria c.1900–c.1919", 2011, 198.

34 *Trade Marks Journal* 48, 1923, 567–568, 2428–2429,

MAP 1.1 *West Africa ca. 1899 before the British occupation of Sokoto*
THE CENTURY ATLAS OF THE WORLD, 1899, NO. 109

and Eastern Trade Corporation, Limited. The latter merged in 1929 with the Niger Company, owned by Lever Brothers, creating a new conglomerate, The United Africa Company Limited (UAC).[35] After this competition reducing amalgamation, and the economic changes wrought by the Great Depression, interest in selling paper waned and the paper trade marks were not renewed.

4 Shifting Trade Routes

There were substantial changes in the political landscape of West and North Africa during the nineteenth century that negatively impacted the traditional, centuries old trans-Saharan trade routes. These routes were stretches of shifting unmarked paths across massive expanses of desert and mountains, a patchwork of drifting trails and footpaths, with multiple routes shifting as political, economic or environmental conditions required. However, during the nineteenth century there were main routes to the Central Sudan: Fez-Taodeni-Timbuktu; Tripoli-Fezzan-Borno that sometimes went through Ghadames and Tripoli-Ghadames-Air-Katsina-Kano, the most important route to Northern Nigeria during this period. In addition, there was abundant internal trade within West Africa. Long distance relay trade was organized and financed by special merchants, with Berbers and Tuareg providing guidance and escort for the camel caravans used to transport merchandise. These trade caravans could take three to six months to cross the Sahara.[36]

The first decade of the nineteenth century saw the Sokoto *ǧihād* and emergence of the Sokoto Caliphate and its take-over of the traditional Hausa states and its great trading cities of Kano and Katsina. Borno's capital, Birni Gazargamu, was sacked twice in 1808 and 1809. This set in motion the collapse of Borno's thousand-year-old Sayfawa dynasty, el-Kanemi's subsequent rise to power and his relocation of Borno's capital to Kukawa in 1814. The French completed its conquest of Algeria by 1848. Britain established colonial rule in Lagos in 1861. France occupied Tunisia in 1881 and Timbuktu in 1894, and Britain militarily occupied Egypt in 1882. The Berlin Conference of 1884 attempted to regulate colonization and trade in the 'scramble for Africa' by European nations. Britain continued its colonialization of 'Nigeria' with occupation of Yoruba kingdoms in 1886, Ijebu kingdoms in 1893 and completed its conquest

35 *The Economist* 3 July 1920, 11; 17 July 1920, 109–110; 30 March 1929, 695–696.

36 S. Baier, *A History of the Sahara in the Nineteenth Century*, 1978; A. B. Sani, "Trade and Diplomatic Relations between the Sokoto Caliphate, Ghadames, Tuat and Abzin in the Nineteenth Century", 2006.

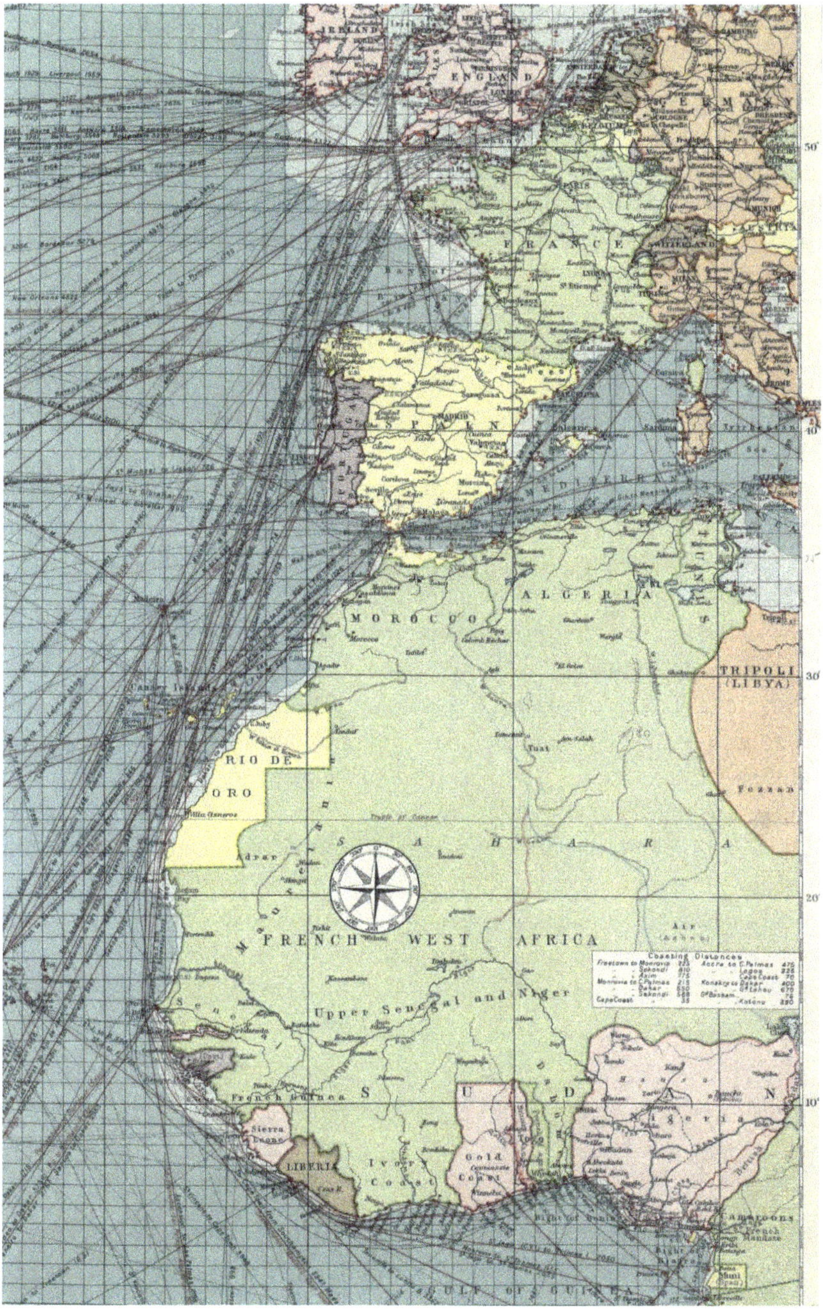

MAP 1.2 *West African shipping lanes ca. 1922*
PHILIPS' MERCANTILE MARINE ATLAS, 1922, MAP I

of Northern Nigeria with the fall of Kano in 1903. Italy declared its suzerainty over Libya in 1911.[37]

There were also significant changes in the efficiency and speed of ocean transport. By the mid-nineteenth century the distance from Liverpool to Lagos, the 'Liverpool' of West Africa, was covered in thirty-five days; by 1900 it took nineteen days. A railway line connecting Lagos with Kano was completed in 1911. The leading firm in the West African shipping business was Elder Dempster of Liverpool with a major office in Lagos.[38] Dozens of relatively small UK trading companies, the majority with headquarters in Liverpool, were doing business in Lagos by the 1880s.[39] British caravan tolls on the trans-Saharan trade increased seven fold between 1904 and 1908 and by 1910 trade across the desert was "practically extinct".[40] In 1917, British naval staff recorded a "sharp fall" in caravan traffic to and from Tripoli with "goods now reaching Kano proceed(ing) by rail to Lagos and are shipped thence to Liverpool".[41] However, some traditional routes persisted as evidenced by the fact that the *SÙRÙ LÓGÙN AIYE* watermark is found in Sokoto[42] and Malian manuscripts. This paper was brought from Liverpool to Lagos by steamship and then, after 1911, by rail to Kano. At this point the paper traveled the old caravan routes to Sokoto and then on to Timbuktu.

With the British in control of Nigeria, and a political stability not seen for more than a century, the paper trade would no longer be dominated by the hand-made Italian papers that had previously been the norm in exports to Northern Nigerian. Machine-made papers, the majority produced in Lancashire and traded through Liverpool—watermarked with Crown Standard; Cranston Ledger; Beale, Ball & Co.; and others—became increasingly common, joining Italian machine-made papers watermarked with *Andrea Galvani Pordenone*; *SSB* (Italiana Cartiera Societa Anonima Serravalle-Sesia-Borgo); *Cartiera gia Lustig*; as well as papers watermarked *Beniamino Arbib*, *A. Ydlibi* and *SÙRÙ LÓGÙN AIYE*.

37 M. E. Chamberlain, *The Scramble for Africa*, 2010; M. Hiskett, *The Sword of Truth*, 1994.

38 A. Olukoju, *The "Liverpool" of West Africa*, 2004, 19.

39 B. W. Hodder, "The growth of trade at Lagos (Nigeria)", 1959.

40 S. Bello, *Sate and Economy in Kano c.1894–1960: a study in colonial domination*, 2011, 93.

41 Great Britain Admiralty, *A Handbook of Libya*, 1917, 101.

42 Personal communication 30 May 2013 from Dr. Sahabi Maidamma Jabo, Principal Librarian, Usman Danfodio University, Sokoto.

5 Conclusion: Paper Use and Dating

In 1869–1870 Nachtigal recorded that in Bornu a *rizma* (ream) of 500 sheets was sold for about six Maria Theresa dollars, the same price as for an old woman slave, and that single sheets could be used in lieu of small change. Reflecting on the paper trade in the early twentieth century, with no mention of scarcity, Malam Koki of Kano recounts that paper was sold by the *ruzuma*, a bundle of five hundred, for twelve to thirteen shillings. Translated into 1913 silver grams the price differential is relatively little but paper was more plentiful in the twentieth century than in earlier times. In Britain alone paper production increased five-fold between 1860 and 1900.[43] Machine-made paper could be produced with identical watermarks made by dandy rolls, in large quantities, and therefore could be purchased in quantity rather than by a few sheets. Paper would therefore not necessarily be used immediately on purchase. With these issues in mind we should be extremely cautious in assigning absolute dates to manuscripts with watermarks made by a dandy-roll on machine-made paper.

Bibliography

Arbib, Beniamino. 1904. *Libro giallo relativo alle elezioni del Collegio del Commercio d'Importazione de 7 e 9 marzo 1904 in Alessandria-Egitto*, Alexandria, E. Sonnino.

Abbott, Nabia. 1949. "Maghribī Koran manuscripts of the seventeenth to the eighteenth centuries", *American Journal of Semitic Languages and Literatures* 55/1, 61–65.

Baier, Stephen. 1978. *A History of the Sahara in the Nineteenth Century*, Working Papers in African Studies, WP4, Boston, Boston University, African Studies Center.

Balboni, Luigi Antonio. 1906. *Gl'Italiani nella civilta egiziana del secolo XIX*, Alexandria, Tipo-litografico via Penasson, vol. 2.

Bargery, George Percival. 1934. *A Hausa-English dictionary and English-Hausa vocabulary*, Oxford, Oxford University Press.

Bello, Sule. 2011. *State and Economy in Kano c.1894 to 1960: a study in colonial domination*, Zaria (Nigeria), Ahmadu Bello University Press.

Biddle, Michaelle. 2008. *Saving Nigeria's Islamic manuscript heritage*, http://works.bepress.com/mbiddle/1/.

43 G. Nachtigal, *Sahara and Sudan. II: Kawar, Bornu, Kanem, Borku, Ennedi*, 1980, 225 and 232); N. Skinner, *Alhaji Mahmudu Koki: Kano malam*, 1977, 33–34; Director of the Mint, *Monetary systems of the principal countries of the world*, 1913, 3, 20; A. H. Shorter, *Paper making*, 1971, 105.

Biddle, Michaelle. 2017. "New strategies in using watermarks to date Sub-Saharan Islamic Manuscripts", in: Andrea Brigaglia & Mario Nobili (eds.), *Arts and Crafts of Literacy in Sub-Saharan Africa*, Berlin, Walter de Gruyter.

Bivar, Adrian David Hugh. 1968. "The Arabic calligraphy of Nigeria", *African Language Review* 7, 3–15.

Blair, Sheila. 2006. *Islamic calligraphy*, Edinburgh, Edinburgh University Press.

Calico Printers' Association, Ltd. v. Barclays Bank, Ltd. and Anglo-Palestine Company, Ltd., King's Bench Division, KBD Oct. 27–30, 1930; Nov. 14, 1930. http://www.docstoc.com/docs/92186770/Calico-Printers-v-barclays accessed 4/1/2013.

Chamberlain, Muriel Evelyn. 2010. *The Scramble for Africa*, Harlow (England), Longman Pearson Education Limited.

Dalla Vedova, Giuseppe. September 1885. "Pellegrino Matteucci ed il sul diario inedito", *Bolletino della Societa Geografica Italiana*, 641–673.

Dillont, A. Price. 1916. "Unusual developments in watermarking methods", *The American stationer* 79 (June), 24–36.

Director of the Mint. 1913. *Monetary systems of the principal countries of the world.* Treasury Department Document no. 2657, Washington, Government Printing Office.

Douglas, Cecil L. 1970. "New concepts in dandy roll design and construction for quality watermarking", *Paper maker and British Paper Trade Journal* 160, 126–129.

Editor of "The Paper Mills Directory". 1874. *The art of papermaking: a guide to theory and practice*, London, Kent & Co.

Export merchant shippers of London, Manchester, Liverpool, Birmingham, Wolverhampton and Walsall, with their respective trading ports, and the class of goods they customarily ship. 1877. London, Dean & Son.

The export merchant shippers of London, Manchester, Liverpool, Birmingham, Wolverhampton, Walsall, Leeds, Bradford, Sheffield, Glasgow, Greenock, Edinburgh, Leith, Hull, Newcastle-on-Tyne, Stockton-on-Tees, Middlesborough, West Hartlepool, Gateshead, Sunderland, North and South Shields, Bristol and Cardiff with their respective trading ports and the class of goods they customarily ship. 1882. London, Dean & Son.

Find A Grave. http://www.findagrave.com/cgi-bin/fg.cgi?page=gr&GRid=52258940 (accessed 4/15/2013).

Gacek, Adam. 2009. *Arabic Manuscripts: A Vademecum for Readers*, Leiden/Boston, E.J. Brill.

Great Britain Admiralty. 1917. *A handbook of Libya*, London, H.M. Stationary Office.

Hackelman, Charles W. 1921. *Commercial engraving and printing: a manual of practical instruction and reference*, Indianapolis, Indiana, Commercial Engraving Publishing Company.

Hartford Courant (Connecticut). 22 September 1954. "Helen Zamulko, Gordon Ydlibi plan to marry", 12.

Hiskett, Mervyn. 1994. *The Sword of Truth*, Evanston (Illinois), Northwestern University Press.

Hodder, Branwell William. October 1959. "The growth of trade at Lagos (Nigeria)", *Tijdschrift voor economische en sociale geografie* 50, 197–202.

Hunter, Dard. 1947. *Papermaking: the History and Technique of an Ancient Craft*, New York, Alfred A. Knopf.

Hunwick, John Owen. 1995. *Arabic Literature of Africa*. Vol. 2: *The writings of Central Sudanic Africa*, Leiden/New York/Köln, E.J. Brill.

King's Bench Division. KBD Oct. 27–30, 1930; Nov. 14, 1930, http://www.docstoc.com/docs/92186770/Calico-Printers-v-barclays (accessed 4/1/2013).

Labarre, Emile Joseph. 1952. *Dictionary and encyclopaedia of paper and paper-making*, Amsterdam, Swets & Zeitlinger.

London Gazette. 4 November 1892. 6192 Beniamino Arbib Clemente Hassan, http://www.london-gazette.co.uk/issues/26341/pages/6192/page.pdf (accessed 4/1/2013).

London Gazette. 1 February 1927. 688 A. Namani A. Ydlibi Azaat Ydlibi, http://www.london-gazette.co.uk/issues/33244/pages/688/page.pdf (accessed 4/1/2013).

London Gazette. 19 December 1967. 13956 M. & G. Ydlibi, http://www.london-gazette.co.uk/issues/44478/pages/13956/page.pdf (accessed 2/1/2014).

Marine Engineer, a monthly journal. 1888–1889. Vol. 10, 394.

Modibbo, Mahmud A. 2011. "The colonial state and European trading companies in Northern Nigeria c.1900–c.1919", in: Musa Adamu Mamman et al. (eds.), *History Department Seminar Series*. Vol. 3: 1978/1979, Session, Zaria (Nigeria), Ahmadu Bello University Press, 184–212.

Nachtigal, Gustav. 1881 (1980). *Sahara and Sudan*. Vol. 2: *Kawar, Bornu, Kanem, Borku, Ennedi*, A. G. B. Fisher & H. J. Fisher (trans.), London, C. Hurst and Company.

Nigerian National Archives Enugu—Archives of Eastern Nigeria branch http://archiveswiki.historians.org/index.php/National_Archives_of_Nigeria (accessed 4/15/2013).

Olukoju, Ayodeji. 2004. *The "Liverpool" of West Africa: the dynamics and impact of maritime trade in Lagos, 1900–1950*, Trenton, New Jersey, Africa World Press, Inc.

Paper Mills Directory of England, Scotland and Ireland for 1871. 1871. London, Kent and Co.

Papermills directory for England, Scotland & Ireland and year book of the paper-making trade for 1882. 1882. London, Kent and Co.

Pettinato, Giovani. 1991. *Ebla, a new look at history*, C. Faith Richardson (trans.), Baltimore, John Hopkins University Press.

Philips' Mercantile Marine Atlas. 1922. London, G. Philip & Son; Liverpool, Philip, Son & Nephew.

Phillips Paper Trade Directory of the World. 1906/1907–1946/1947. London, S. C. Phillips & Co.

Robinson, Charles Henry. 1913. *Dictionary of the Hausa Language*. Vol. 1: *Hausa-English*, Cambridge, University Press.

Safwat, Nabil F., with a contribution by Mohamed Zakariya. 1996. *The art of the pen: calligraphy of the 14th to 20th centuries*, London, The Nour Foundation in association with Azimuth Editions, Oxford University Press.

Safwat, Nabil F. 2000. *Golden pages: Qur'ans and other manuscripts from the collection of Ghassan I. Shaker*, Oxford, Oxford University Press for Azimuth Editions.

Sani, Abubakar B. 2006. "Trade and Diplomatic Relations between the Sokoto Caliphate, Ghadames, Tuat and Abzin in the Nineteenth Century", in: Hamid Bobboyi & Mahmood Yakubu Alhaji (eds.), *The Sokoto Caliphate: History and Legacies, 1804–2004*, Kaduna, Arewa House, Ahmadu Bello University, vol. 1, 253–268.

Senate. 1902. *Annual Report of the Commission of Patents for the Year 1901*, 57th Congress First Session, S. Doc.151, Washington, Government Printing Office.

Shenton, Robert. 2011. "A note on the origins of European commerce in Northern Nigeria", in: Musa Adamu Mamman et al. (eds.), *History Department Seminar Series*. Vol. 1: 1976/1977 Session, Zaria (Nigeria), Ahmadu Bello University Press, 166–175.

Shorter, Alfred Henry. 1971. *Paper making in the British Isles: an historical and geographical study*, Newton Abbott (Devon), David and Charles.

Sindall, Robert Walter. 1920. *Paper Technology: an Elementary Manual on the Manufacture, Physical Qualities and Chemical Constituents of Paper and of Papermaking Fibres*, London, Charles Griffin & Company, Limited.

Skinner, Neil. 1977. *Alhaji Mahmudu Koki: Kano malam*, Zaria (Nigeria), Ahmadu Bello University Press.

Smith, Benjamin Eli. 1899. *The Century Atlas of the World*, New York, The Century Co.

Smith, Larry D. 2003. "The paper machine", *Watermarks: a brief history of watermarks*. http://www.motherbedford.com/watermarks/Watermark1B8.htm (accessed 6/29/2013).

Spicer, A. Dykes. 1907. *The Paper Trade a descriptive and historical survey of the paper trade from the commencement of the nineteenth century*, London, Methuen & Co.

Steamer: City of Birmingham. 1917. Nature of goods: Cotton Goods. National Archives U.K. Reference BT 365/4/132, Azaat Ydlibi and Company 8/02/1917, http://discovery.nationalarchives.gov.uk/SearchUI/Details?uri=C13357762 (accessed 4/1/2013).

Street's Indian and colonial mercantile directory for 1870. 1870. London, G. Street and Street Brothers.

Tomlinson, Charles (ed.). 1855. "Paper", *Cyclopaedia of useful arts and manufactures*, London, James S. Virtue, 357–374.

Trade Marks Journal. 1876–1932. London, Patent Office.

United States Department of Commerce Bureau of Foreign and Domestic Commerce. 1925. *Paper and Paper Products in the Union of South Africa and Egypt*, Trade Information Bulletin 363.

Walz, Terence. 1985. "The paper trade of Egypt and the Sudan in the Eighteenth and Nineteenth Centuries", in: Martin William Daly (ed.), *Modernization in the Sudan: Essays in Honor of Richard Hill*, New York, Lillian Barber Press, Inc., 29–48.

Walz, Terence. 2011. "The paper trade of Egypt and the Sudan", in: Graziano Krätli & Ghislaine Lydon (eds.), *The Trans-Saharan Book Trade: Manuscript culture, Arabic literacy and Intellectual History in Muslim Africa*, Leiden, E. J. Brill, 73–107.

Wellisch, Hans H. 1981. "Ebla: The World's Oldest Library", *The Journal of Library History* 16, 488–500.

World's paper trade review. 1883–1908. London, Stonhill & Gillis.

WRECKsite. http://www.wrecksite.eu/wreck.aspx?155797 (accessed 4/15/2013).

Photographs

Photographs are by the author and were made using transmitted light and a Canon S100 camera.

CHAPTER 2

Note on a Dated Tunisian Watermark (1860–1861)

Michaelle Biddle

Résumé

La collection 3 de manuscrits islamiques, 653, conservée à la section « Manuscrits » du Département des livres rares de la Bibliothèque de l'Université de Princeton, se compose d'une série de documents du gouvernement tunisien, datés de 1277/1860–1861. Ils ont été rédigés sur du papier vélin à la marque du Ministre des Affaires étrangères, Muṣṭafā Ḫaznadār, en arabe – *wizārat al-ḫariǧiyya Muṣṭafā Ḫaznadār*.

Manuscripts Division, Department of Rare Books and Special Collections, Princeton University Library, Islamic Manuscripts Third Series 653 is comprised of a collection of Tunisian government documents, including the first written constitution in the Middle East. According to the colophon these documents were written in Muḥarram 1277/July 1860 at the Bardo Palace in Tunis at the command of Muḥammad al-Ṣādiḳ Bey (1814–1882), the twelfth ruler of the Husainid Dynasty. Two of the documents are dated Rağab 1277/February 1861.[1] Instead of official paper with a printed letterhead, watermarked paper was used. The bulk of the documents concern internal affairs—regulations concerning Jewish representation, regulations concerning various judicial courts, the constitution, yet the watermark indicates the foreign/external ministry.

These documents are on burnished, ivory white, wove paper. Whilst during the 19th century Italians were the largest foreign population in Tunisia, and had for centuries extensively traded paper to Africa, during the *risorgimento* Italian paper production was not robust.[2] Given the dominance of the French during the period when these documents were written, the paper is probably French. The paper has characteristics of mid-nineteenth century French paper—soft, highly transparent, ivory white, smooth with an even formation.

1 Princeton University Library Main Catalog (OCLC) 607557560 (accessed 16 June 2015).

2 A. Pellegrini, *L'industria della carta in Italia*, 1954, 91–93.

 | DOI 10.1163/9789004360877_004

FIGURE 2.1 *Watermark in Princeton University Library, Islamic Manuscripts, Third Series, no. 653. Watermark 81 × 38 mm*
PHOTO BY THE AUTHOR

It bears similarities with many of the papers in Imprimerie Royale imprints, including that of *Exploration scientifique de l'Algérie* (ESA).[3]

Eight of the one hundred twelve sheets are watermarked in Arabic script—*wizārat al-ḫariǧiyya Muṣṭafā Ḫaznadār*. The first line can be read as 'foreign (external) ministry of', though in Arabic the correct spelling is *ḫāriǧiyya.* The second line is the name 'Muṣṭafā Ḫaznadār'. Ḫaznadār is the Arabic version of the Ottoman Turkish (Osmânî) title Hazinedar and means 'treasurer'.[4]

Muṣṭafā Ḫaznadār (original name Giorgios (Stephanou) Kalkias Stravelakis, born 1817 Kardamila, Chios, died 26 July 1878 Tunis, Tunisia) was a Greek, captured as a five-year old after the Chios massacre and sold as a slave (*mamlūk*) by the Ottomans to an envoy of the Husainids, themselves of Muslim Greek Cretan origin. The Husainids ruled Tunisia—the Beylik of Tunis—under the suzerainity of the Ottoman Empire from 1705. Ḫaznadār is a *laqab* (title and

3 Louis André of Rennes University has sampled the voluminous Imprimerie Royale archives 1829/39/49 and found that in 1839 there were fifteen suppliers of paper, for the most part the large wholesale paper merchants of Paris who obtained paper from various paper mills all over France. My thanks to Claire Bustarret for facilitating this information.

4 My thanks to Anne Regourd, Davison MacLaren, Steven Blackburn and M'hamed Oualdi for comments and assistance with the intricacies of this translation.

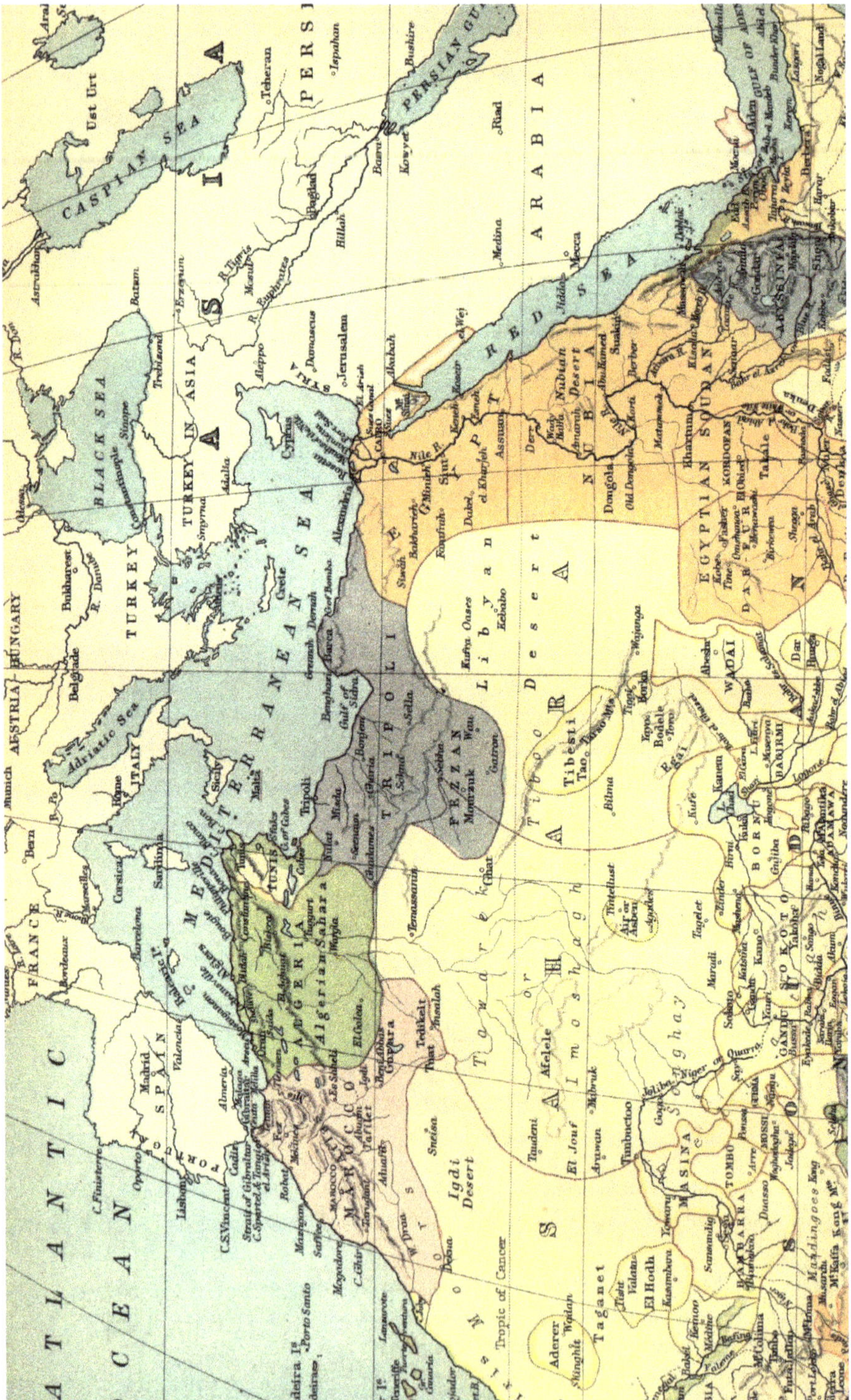

MAP 2.1 *North Africa in 1885*
THE SCOTTISH GEOGRAPHICAL MAGAZINE 1, 1885

part of a name) that Muṣṭafā received like other mamluks of his kind. Ḫaznadār converted to Islam and was raised by the Husainid family. He married into the family and governed as *Wazīr al-awwāl* (prime minister) from 1837 to 1873. In 1864 he attempted to squeeze more taxes from the peasantry in order to pay off the crippling foreign debts. The countryside rose in revolt, brutally supressed by the venal Ḫaznadār, and the constitution was abandoned. He fell from power in 1873 while attempting to play the country's creditors against one another.[5]

The French occupied Algeria in 1830 and for decades thereafter aggressively looked south and east to expand their investment opportunities. For example in 1844 Ernest Carette (1808–1890), a Saint-Simonian army officer and later prefect of Constantine, published comprehensive and meticulous data on the Algerian desert and its commerce gathered from native Algerian or Tunisian merchant informants, in the multivolume compilation *Exploration scientifique de l'Algérie* (ESA), modelled on Napoleon's *Description de l'Égypte*.[6] Commerce was regarded by the French military as a necessary adjunct to political and strategic domination.

Ill-conceived Tunisian development projects, corruption, rapacious lending by European banks and subsequent inability to repay foreign loans paved the way for the French invasion of 1881 when Tunisia became a protectorate.[7]

Bibliography

Carette, Ernest. 1844. *Recherches sur la géographie et le commerce de l'Algérie méridionale* (*Exploration scientifique de l'Algérie II*), Paris, Imprimerie Royale.

Cherif, Mohamed Hedi. 1989. "New trends in the Maghrib: Algeria, Tunisia and Libya", in: Jacob Festus Ade(niyi) Ajayi (ed.), *Africa in the nineteenth Century until the 1880s, General History of Africa VI*, Berkeley, University of California Press & UNESCO, 448–477.

Oualdi, M'hamed. 2011. *Esclaves et maîtres: les mamelouks des beys de Tunis du XVIIe siècle aux années 1880*, Paris, Publications de la Sorbonne.

Perkins, Kenneth J. 2004. *A history of modern Tunisia*, Cambridge, Cambridge University Press.

Pellegrini, Andrea. 1954. *L'industria della carta in Italia*, Rome, Pubblicazioni dell'ente Nazionale per la Cellulosa e per la Carta.

5 K. Perkins, *A History of Modern Tunisia*, 2004, 10–38; M. Oualdi, *Esclaves et maîtres*, 2011.

6 E. Carette, *Recherches sur la géographie et le commerce de l'Algérie méridionale*, 1844.

7 M. Cherif, "New trends in the Maghrib: Algeria, Tunisia and Libya", 1989, 455–471.

CHAPTER 3

Recalling *Alikurna*: "ليكوريا" Countermarked Paper among Scribes in the Late 19th Century Ottoman Levant

Evyn Kropf

Résumé

Le terme « Alikurna » a été employé par les Ottomans dans leur classification des papiers aussi tôt que le milieu du xviie s. On pense qu'il a son origine dans le nom corrompu d'une ville – ou d'une région – italienne, fournisseur de ce type de papier, plus exactement Livourne ou la Ligurie. Au xixe s., les producteurs de papier italiens rappellent cette dénomination par une contremarque en écriture arabe « ليكوريا » placée dans leurs papiers destinés au marché ottoman. L'un de ces papiers portant la marque du fabricant Ferdinando Betti se trouve dans un manuscrit actuellement conservé à la Bibliothèque de l'Université du Michigan. Contenant l'histoire en arabe d'un héros populaire en 25 petits volumes, le manuscrit a été copié en 1895 par Muḥammad b. Sulaymān de la famille bien connue des Nablusi, ʿArafāt. Mais le même papier (de même que d'autres) apparaît également dans des manuscrits samaritains contemporains de la région de Nablus, ainsi que dans des manuscrits en arabe provenant de Damas. Prenant la notion de « papier Alikurna » comme point de départ, cet article explore les origines possibles de cette contremarque en arabe nouvellement identifiée et s'intéresse au sens de son apparition dans des produits italiens à partir de l'étude de la production italienne et de sa circulation dans le milieu des scribes du Levant ottoman, à la fin du xixe s.

1 Introduction

Over the centuries, a vast array of watermarked paper circulated among scribes in the Ottoman lands. Most of these watermarked papers were of European manufacture, with firms of particular regions dominating the markets at different times.[1]

1 Cf. V. Nikolaev, *Watermarks of the Mediaeval Ottoman Documents in Bulgarian Libraries*, 1954, vol. 1, 13–15, 63–64', 123–124', 469–470.

 | DOI 10.1163/9789004360877_005

Papermaking firms typically chose emblems representative of the production in their regions. Nevertheless, imitation of emblems among papermakers across regions was not uncommon. In addition to an emblem or motif, many watermarks incorporated words and initials intended to more specifically identify the paper's origins. In this form of branding, initials or a full spelling of the name of the papermaking firm were provided, sometimes accompanied by the name of the city or region where the firm was located and even the date of manufacture. These identifying countermarks became quite common among European firms and were subject to regulation.[2]

At some stage, the Ottomans began applying the term *Alikurna* to these watermarked imports.[3] The term is believed to have been derived from the corrupted name of the Italian city or region supplying such paper, possibly Livorno or Liguria, and documentary evidence for its use as a designation of paper type appears as early as the mid-17th century.[4]

Compared with Latin and Cyrillic scripts, watermarks and countermarks incorporating Arabic script are quite uncommon and virtually unattested prior to the early 19th century.[5] Almost all of the examples known to appear in papers that circulated among the Ottomans have been attributed to Ottoman manufacture. These watermarks use the Arabic script exclusively and in at least two cases include an explicit reference to Istanbul, presumably as place of manufacture, along with a reference to the year of the Hiğrī calendar in Hindu-Arabic numerals, presumably as date of manufacture.[6]

---

2 By 1774, Venetian law required all paper to bear the initials of the papermaker and the date, though this was left to the imagination of the manufacturers. Cf. T. Walz citing Fedrigoni, "The Paper Trade of Egypt and the Sudan in the Eighteenth and Nineteenth Centuries and its Re-export to the Bilad as-Sudan", 2011, 84.

3 Cf. G. Erişen-Yazıcı, ALİKURNA, 2001–, 124, vol. 1.

4 Cf. M. A. Kâğıtçı, *Kağıtçılık Tarihçesi*, 1936, 213–214, in which he cites the investigations of Muallim Cevdet Bey (d. 1935) regarding a listing of paper arrivals for the year 1051/1641 or 2 as follows "1051 (1641) senesinde İstanbula gelen kağıtlar: Hindkari kağıt, Buyrultu için örta battal kağıt, Zarflık kağıt, Telhis kağıdı, Alıkurna kağıdı".

5 For another example, see the case of the Abū Šubbāk papers addressed by Anne Regourd in the present volume and in her "Introduction to the Codicology: Papers, Ruling, and Bindings", 2014, xlix ff.

6 Cf. V. Nikolaev, *Watermarks*, 1954, nos. 938, 944, 945, 946, 947, 977; S. Ünver, "XV. Yüzyılda Türkiye'de Kullanılan Kâğıtlar ve Su Damgaları", 1962, 743, 762; O. Ersoy, *XVIII. ve XIX. Yüzyıllarda Türkiye'de Kâğıt*, 1963, 57, 58, 193, catalogue nos. 4–5; M. A. Kâğıtçı, *Historical Study of Paper Industry in Turkey = Historique de l'industrie papetière en Turquie*, 1976, 11 and Fig. 7; Ş. Tekin, *Eski Türklerde Yazı, Kâğıt, Kitap ve Kâğıt Damgaları*, 1993, 106 referencing "Resim 24".

An exception is a rare 19th century watermark identified by Osman Ersoy in the paper of the court registers of Siirt, dated 1288–1292/1871–1875 and held now in the Diyarbakır Museum.[7] This mark incorporates an Arabic script countermark together with the name of the Italian firm Fratelli Palazzuoli in Latin characters.

I have since identified another such Arabic script countermark alongside the name of the Italian papermaker Ferdinando Betti in a paper used in a manuscript now held at the University of Michigan Library. Carrying the text of an Arabic popular heroic narrative in 25 slim volumes, the manuscript was copied in 1895 by Muḥammad b. Sulaymān ʿArafāt. Intriguingly, this same paper has also been identified in contemporary Samaritan manuscripts from the Nablus region and in one contemporary Arabic manuscript likely of Damascene provenance.[8]

As an apparent Arabic script rendering of the name of the Italian city Liguria—that is, "ليكوريا"—the countermark of this paper recalls the Alikurna paper apparently known to the 17th century Ottomans and suggests the use of a recognizable designation by 19th century Italian papermakers intending to circulate their product in Ottoman markets.

Taking the notion of Alikurna paper as a point of departure, this article explores the possible origins of this newly identified Arabic script countermark and the significance of its appearance in an Italian product by examining the evidence for its Italian production and its circulation in scribal communities of the late 19th century Ottoman Levant.

2 Alikurna Paper

It is clear from limited documentary evidence that the term Alikurna (or Ali Kurna, from عليقورنه) circulated among the Ottomans as a designation of paper type. However, the extent and significance of the circulation and indeed

7 O. Ersoy, *XVIII. ve XIX. Yüzyıllarda Türkiye'de Kâğıt*, 1963, 21 (note 5), 56, insert between 68–69, and 194.

8 These include papers of two Samaritan manuscripts from the British Library in which I have been able to observe and identify the mark, as well as papers of several Samaritan manuscripts now held at the Bibliothèque nationale de France (BnF), in which Jean-Pierre Rothschild apparently observed this Ferdinando Betti mark though without full identification of the countermark, cf. his *Catalogue des manuscrits samaritains*, 1985, 21. I have not yet been able to examine any of the BnF manuscripts to confirm the identification of the countermark.

even the origins of the term remain obscure. Mehmet Ali Kâğıtçı's reference to a listing of paper arrivals for the year 1051/1641–1642 presumably obtained from Muallim Cevdet Bey represents the earliest use of the Alikurna designation thus far encountered.[9] Other uses of the term appear as a direct reference to a place name. An 18th century document cited by Ahmet Refik includes such a use, which he interprets as a reference to "Liguri".[10] Still another document dated Ǧumādā II 1259/June–July 1843 and pertaining to the establishment of the paper mill at Izmir mentions the use of several types of paper in Istanbul at that time, not only from London but from Trieste and "عليقورنه" (Alikurna).[11]

According to Süheyl Ünver, the term represents a corruption of the name of the Italian city Livorno ("Elikorne" or Leghorn), an important port for the export of Italian goods, including paper. He argues that even papers manufactured in other cities and regions—such as Venice and Central Europe—were exported through this port, which, from the perspective of the Ottomans, became the *de facto* point of origin for European papers of various types. Thus, various watermarked papers irrespective of place of manufacture were referenced among the Ottomans by the same term derived from the name of this port, i.e. "Ali Kurna".[12]

Osman Ersoy argues that it seems more likely that the term Alikurna is not derived from the name Livorno, but from Liguria. He bases this argument on the aforementioned reference to the term in the work of Ahmet Refik, in which Refik identifies Alikurna as "Liguri" without further explanation or controversy, and on Ersoy's own reading of an intriguing watermark appearing in the paper of the legal registers of Siirt, dated 1288–1292/1871–1875 and held now in the Diyarbakır Museum. He reproduces the mark in his catalogue at number 35.[13,14] [Fig. 3.1]

The large mark includes a name in Latin script, FRATELLI PALAZZUOLI—that is, the Palazzuoli Brothers, presumably an Italian firm—above a

---

9 Again, cf. M. A. Kâğıtçı, *Kağıtçılık Tarihçesi*, 1936, 213–214; on Muallim Cevdet Bey cf. A. G. Sayar, MUALLİM CEVDET, 1988–, vol. 30, 314.

10 Document 232 "Muhasım devletlerin birbirinden zabt iderek Türkiye limanlarına getirdikleri eşyadan gümrük resmi alınacağına dair", dated mid Raǧab 1173/Feb.–Mar. 1760 in A. Refik, *Hicrî On İkinci Asırda İstanbul Hayatı (1100–1200)*, 1930, vol. 3, 191–193 (1988 reprint).

11 Cf. "*İBDA, Cevdet-İktisat*, No. 865 (Cemaziyülahir, 1259)" cited in O. Ersoy, *XVIII. ve XIX. Yüzyıllarda Türkiye'de Kâğıt*, 1963, 21, 25, 48 ff.

12 Cf. S. Ünver, "XV. Yüzyılda Türkiye'de Kullanılan Kâğıtlar ve Su Damgaları", 1962, 740–741.

13 Cf. O. Ersoy, *XVIII. ve XIX. Yüzyıllarda Türkiye'de Kâğıt*, 1963, 21, note 4.

14 Cf. O. Ersoy, *XVIII. ve XIX. Yüzyıllarda Türkiye'de Kâğıt*, 1963, 21, note 5, 56, insert between 68–69, and 194 (index of watermarks).

FIGURE 3.1 *Fratelli Palazzuoli mark*
REPRODUCED IN OSMAN ERSOY, XVIII. VE XIX. YÜZYILLARDA TÜRKİYE'DE KÂĞIT, CATALOGUE NO. 35

countermark in unpointed Arabic script, the word "لكوريا".[15] Ersoy has read this countermark as a phonetic spelling for the name Liguria, that is "ليكوريا". Nevertheless, the unpointed letters mean that other readings are also possible, including "ليكورنا", which phonetically would recall the English form of Livorno—Leghorn. The latter reading would of course favor the identification proposed by Süheyl Ünver.

Another instance of this countermark could potentially clarify the reading, and indeed I have been able to identify another version in a paper used in a manuscript now held at the University of Michigan Library, Isl. Ms. 653, vols. 1–25, copied in 1895. In this case, the countermark appears under the name FERDINANDO BETTI—presumably another Italian papermaker—which itself appears under a version of the popular *Tre lune* or three crescents motif. Unlike the Palazzuoli countermark cited by Ersoy, in this case the Arabic script of the countermark is pointed to read "ليكوريا" for a phonetic spelling of the name Liguria. However, only two dots are provided and they are placed somewhat unexpectedly, even haphazardly, perhaps betraying a poor knowledge of the language. [Fig. 3.2]

Intriguingly, another watermarked paper bearing a mark of the Palazzuoli Brothers also appears among the papers of this manuscript, and evidence that

15 I have not had the privilege of examining this paper, but Ersoy seems to have insisted on the unpointed nature of the Arabic script in the countermark, reproducing it in this fashion in both his text and his catalogue (cf. O. Ersoy, *XVIII. ve XIX. Yüzyıllarda Türkiye'de Kâğıt*, 1963, 21, 68–69).

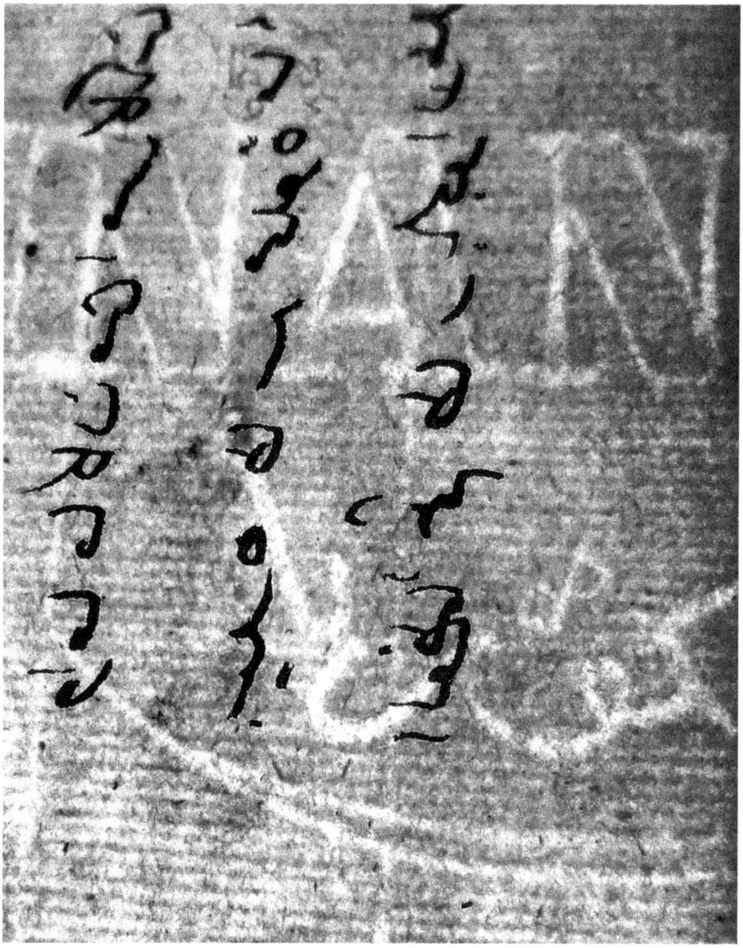

FIGURE 3.2 *View of the pointing in the "ليكوريا" Arabic script countermark in Isl. Ms. 653, vol. 24, pp. 3/4 (reversed, transmitted light image, transformed into the RGB colorspace with B channel isolated by RetroReveal.org)*

these papers circulated closely, even together, exists in other manuscripts.[16] [Tables 1 & 2].

Certainly with both the Palazzuoli and Ferdinando Betti watermarked papers, the intention with the Arabic script countermark seems to be a reference to an Italian place name, specifically one that can be linked to the 17th century Ottoman Alikurna designation, either Liguria or Livorno.

16 For further discussion of 19th century circulation of Palazzuoli marked paper, see the Appendix.

TABLE 1 *Ferdinando Betti/"ليكوريا" watermarked papers*

Shelf mark	Folio Dimensions (mm.)	Date	Watermark	Laid and Chain Line Measurements
Isl. Ms. 653 v. 3–v. 25 (University of Michigan)	224 × 165	ca. 20 Ḏū al-Ḥiǧǧa 1312 [ca. 14 June 1895]	*Tre lune* above FERDINANDO BETTI ليكوريا	9–10 laid lines per cm. (v) chain lines spaced about 23–26 mm. apart (H)
British Library Or. 10370	162 × 113	25 Ǧumād I 1316 [ca. 11 Oct 1898]	*Tre lune* above FERDINANDO BETTI ليكوريا	9 laid lines per cm. (H) chain lines spaced ~21 mm. apart (v)
British Library Or. 10723	214 × 155	undated	*Tre lune* above FERDINANDO BETTI ليكوريا	9 laid lines per cm. (H) chain lines spaced ~21 mm. apart (v)
BnF sam. 33	212 × 166	"27 Rağab 1320 H./30 octobre 1902"	« trelune », monogramme arabe et nom de fabricant: « FERDINANDO BETTI »	(unreported)
BnF sam. 37	206–207 × 157	undated	« trelune », monogramme arabe et nom de fabricant: « FERDINANDO BETTI »	(unreported)

Tre lune	Copyist/Provenance	Citations	Notes
78 mm. long, perpendicular to chains	Muḥammad ibn al-Ḥāğğ Sulaymān ʿArafāt Nablus or Damascus?	Online catalogue records, University of Michigan Library	v. 4 also *FF. PALAZZUOLI* with *Tre lune* paper (see Table 3) v. 24 also *Tre lune* and "S S B" with "B" as point of triangle, chains (H) spaced 25 mm. apart, 12 laid lines per cm. (V)
Measurement impossible due to folding/trimming	on f. 51, Arabic colophon with "تمّ على يد (...) اسحاق اهكهن ٥٢ جماداول سنه ٦١٣ ثلاثمايه وسته عشر عربيه الموافق ٤٢ ايلول روميه" Crown identifies the copyist as "Isaac b. Amram b. Salama b. Tabya" (p. 56)	no. 38, Crown, 55–56	Cf. ff. 39, 40, 51
Measurement impossible due to folding/trimming	عزيز ابن يعقوب ابن يوسف چلبى سفرى (colophons on f. 19a and f. 50a) "Shechem"	no. 109, Crown, 145–146	Visible in f. 31; single bifolium in a codex mainly comprised of types of *FF. PALAZZUOLI*/ *Tre lune* paper (see Table 3)
(unreported)	"Ṭāhor ibn Yaʿqūb ibn Ismāʿīl ibn Ibrahim as-Sarāwī ad-Danafī"	no. 51 Rothschild, 130–133	
(unreported)	"Finḥas ibn Isḥaq"	no. 62 Rothschild, 147–148	

TABLE 1 · *Ferdinando Betti/"ليكوريا" watermarked papers* (cont.)

Shelf mark	Folio Dimensions (mm.)	Date	Watermark	Laid and Chain Line Measurements
BnF sam. 49	219–220 × 162–163	"le 29 Ǧumada [I?] 1312 H." "28 novembre 1894"	« trelune », monogramme arabe et nom de fabricant: « FERDINANDO BETTI »	(unreported)
BnF sam. 50	210 × 157	"le 23 Ǧumada I (1)312" "22 novembre 1894"	« trelune », monogramme arabe et nom de fabricant: « FERDINANDO BETTI »	(unreported)
BnF sam. 53	209 × 160	"29 du l-Qaʿda (1)314 H." "1er mai 1897"	« trelune », monogramme arabe et nom de fabricant: « FERDINANDO BETTI »	(unreported)
BnF sam. 55	212 × 159	"12 du l-Qaʿda suivant [i. e. 1311]" "17 mai 1894"	« trelune », monogramme arabe et nom de fabricant: « FERDINANDO BETTI »	(unreported)
BnF sam. 58	214 × 154	"25 Rabi II (1)314 / 20 Ellul 312" "3 octobre 1896"	« trelune », monogramme arabe et nom de fabricant: « FERDINANDO BETTI »	"pontuseaux espaces de 24 à 26 mm"

Tre lune	Copyist/Provenance	Citations	Notes
(unreported)	"Na'aman b. Yōsef ha-Danafī"	no. 38 Rothschild, 101–103	« trelune », lettres « SSB » formant un triangle dont la pointe du bas est occupée par le B
(unreported)	"'Azīz walad Ya'qūb walad Čalabī ibn Yūsef ibn Halabī ad-Danafī (..). Mūsawī (..). Samarī"	no. 33 Rothschild, 88–90	
(unreported)	"Ḥilmī Ḥalabī Sāmarī pour son frère 'Azīz Ḥalabī" (f. 18)	no. 40 Rothschild, 105–108	second colophon on f. 52 with name of "'Azīz ibn Ya'aqob ibn Halabī ibn Yūsef ibn Ḥalābī ad-Danafī" dated "17 Muḥarram 1317 H./28 mai 1899" pertains to section carrying watermark of *Tre lune* with "S S B" and double chain lines
(unreported)	"Na'mān al-Yūsef"	no. 43 Rothschild, 113–116	
(unreported)	"Ibrahīm, dit Barhūm, ibn Yūsef ibn 'Abd el-Laṭīf ibn Ismā'īl as-Sarāwī ad-Danafī al-Mūsawī al-Isrāīl" [sic]	no. 23 Rothschild, 66–67	all folios except ff. 4–9 and 92–[97]

TABLE 1 *Ferdinando Betti/"ليكوريا" watermarked papers* (cont.)

Shelf mark	Folio Dimensions (mm.)	Date	Watermark	Laid and Chain Line Measurements
BnF sam. 61	213 × 160–162	"15 Rabi (I?) (1)316" "3 août 1898"	« trelune », monogramme arabe et nom de fabricant: « FERDINANDO BETTI »	(unreported)
BnF Ar. 6956	220 × 165	12 šaʿbān 1312 8 février 1895	Papier filigrané. Filigrane en forme de trois croissants au nom de FERDINANDO BETTI avec lettres arabes	(unreported)
CW 2481	315 × 220	1297–1311 [1880–1894]	*Tre lune* with countermark "*FB*"	8 laid lines per cm. (H) chain lines spaced 25–26 mm. apart (v)

Isl. Ms. 653 v. 1–25 Online catalogue record, University of Michigan Library (Evyn Kropf) | https://mirlyn.lib.umich.edu/Record/006805676/Description#tabs.

Crown, Alan David. 1998. *A Catalogue of the Samaritan Manuscripts in the British Library*, London, British Library.

Rothschild, Jean-Pierre. 1985. *Catalogue des manuscrits samaritains*, Paris, Bibliothèque nationale, Département des Manuscrits.

Tre lune	Copyist/Provenance	Citations	Notes
(unreported)	"ʿAzīz ben Yaʿqūb Čalabī ben Yūsef Čalabī (*in marg.*), Danafī"	no. 49 Rothschild, 125–128	
(unreported)	محمد أريب بن أرسلان التقي | "copie achevée dans la journée du jeudi 12 šaʿbān 1312 h./ 8 février 1895 par Muḥammad Arīb ibn al-Sayyid Arslān al-Taqī d'après la main du compilateur, frère de la grand-mère maternelle du copiste (f. 35v–36)"	Arabe 6956. Traité juridique Notice descriptive. Bibliothèque nationale de France, 2008 http://archivesetmanuscrits.bnf.fr/ark:/12148/cc8977v	
~65 mm. long, perpendicular to chains			blue tint; flecks of blue and red threds visible

TABLE 2 *FF. PALAZZUOLI watermarked papers*

This table includes a selection of the FF. PALAZZUOLI watermarked papers attested in Arabic and (Cf. A. D. Crown, *Samaritan Scribes and Manuscripts,* 2001, 106) including BL Or. 2691 (A. D. Crown, ca. 1875). Additionally, Rothschild has identified FF. PALAZZUOLI watermarked paper in at least by the priest Ya'qūb b. Hārūn in 1884, BnF sam. 23 (cf. J.-P. Rothschild, *Catalogue des manuscrits sa-*

Shelfmark	Folio Dimensions (mm.)	Date	Watermark	Laid and Chain Line Measurements
Isl. Ms. 653 v. 1 (University of Michigan)	217 × 160	ca. 20 Ḏū al-Ḥiǧǧa 1312 [ca. 14 Jun. 1895]	*FF. PALAZZUOLI* under *Tre lune*	~ 9 laid lines per cm. (V) and chain lines spaced about 28–31 mm. apart (H)
Isl. Ms. 653 v. 2 (University of Michigan)	228 × 167	ca. 20 Ḏū al-Ḥiǧǧa 1312 [ca. 14 Jun. 1895]	*FF. PALAZZUOLI* under *Tre lune*	~ 9 laid lines per cm. (V) and chain lines spaced about 28–31 mm. apart (H)
Isl. Ms. 653 v. 4 (University of Michigan)	230 × 167	ca. 20 Ḏū al-Ḥiǧǧa 1312 [ca. 14 Jun. 1895]	*FF. PALAZZUOLI* under *Tre lune*	~ 9 laid lines per cm. (V) and chain lines spaced about 28–31 mm. apart (H)
British Library Or. 10723	214 × 155	undated	*FF. PALAZZUOLI* under *Tre lune*	9 laid lines per cm. (V) and chain lines spaced 31 mm. apart (H)
British Library Or. 10723	214 × 155	undated	*FF. PALAZZUOLI* under *Tre lune*	6-7 laid lines per cm. (V) and chain lines spaced 30–31 mm. apart (H)

Samaritan manuscripts. Of course numerous other Samaritan manuscripts carry such watermarks
A Catalogue of the Samaritan Manuscripts in the British Library, 1998, 54–55, no. 37, added leaves
10 Samaritan manuscripts at the Bibliothèque nationale de France, including one manuscript copied
maritains, 1985, 96). The BnF manuscripts identified by Rothschild range in date from 1866 to 1899.

Tre lune	Shape of Period	Orientation of ZZ	Copyist / Provenance	Citations	Notes
67–68 mm. long, perpendicular to chains above and slightly to the left of FF. PAL …	triangle	ZZ	Muḥammad b. al-Ḥāǧǧ Sulaymān ʿArafāt Nablus or Damascus?	Online catalogue record, University of Michigan Library	
67–68 mm. long, perpendicular to chains above and slightly to the left of FF. PAL …	triangle	ZZ	Muḥammad b. al-Ḥāǧǧ Sulaymān ʿArafāt Nablus or Damascus?	Online catalogue record, University of Michigan Library	
67–68 mm. long, perpendicular to chains above and slightly to the left of FF. PAL …	triangle	ZZ	Muḥammad b. al-Ḥāǧǧ Sulaymān ʿArafāt Nablus or Damascus?	Online catalogue record, University of Michigan Library	Also *Tre lune* above *FERDINANDO BETTI* (ليكوريا) paper
~70 mm. long, perpendicular to chains (see f. 4)	triangle (cf. f. 7)	ZZ (cf. ff. 3, 10)	عزيز ابن يعقوب ابن يوسف چلبی سفری (colophons on f. 19a and f. 50a) "Shechem"	no. 109, 145–146 in Crown	Also *Tre lune* above *FERDINANDO BETTI* ليكوريا paper (marks visible in f. 31)
~75 mm. long, perpendicular to chains (see f. 4)	circle	ZZ	عزيز ابن يعقوب ابن يوسف چلبی سفری (colophons on f. 19a and f. 50a) "Shechem"	no. 109, 145–146 in Crown	F. 11 onward

TABLE 2 *FF. PALAZZUOLI watermarked papers* (cont.)

Shelfmark	Folio Dimensions (mm.)	Date	Watermark	Laid and Chain Line Measurements
British Library Or. 7928	212 × 160	1303 [1885 or 6]	*FF. PALAZZUOLI* under *Tre lune*	6 laid lines per cm. (V) and chain lines spaced 28 mm. apart (H)
British Library Or. 7928	212 × 160	1303 [1885 or 6]	*FF. PALAZZUOLI* under *Tre lune*	12 laid lines per cm. (V) and chain lines spaced 30 mm. apart (H)
British Library Or. 10874	215 × 155	25 Ṣafar 1301 [ca. 26 Dec. 1883]	*FF. PALAZZUOLI* under *Tre lune*	6–7 laid lines per cm. (V) and chain lines spaced 30–31 mm. (H)
British Library Or. 10874	215 × 155	25 Ṣafar 1301 [ca. 26 Dec. 1883]	*FF. PALAZZUOLI* under *Tre lune*	6 laid lines per cm. (V) and chain lines spaced 28 mm. apart (H)
British Library Or. 10811	215 × 160	23 Raǧab 1297 [ca. 1 Jul. 1880]	*FF. PALAZZUOLI* under *Tre lune*	6 laid lines per cm. (V) and chain lines spaced 28 mm. apart (H)
British Library Or. 1448	213 × 145	ca. 13 šawwāl 1289 [ca. 14 Dec. 1872]	*FF. PALAZZUOLI* under *Tre lune*	6 laid lines per cm. (V) and chain lines spaced 28 mm. apart (H)
British Library Or. 12257	220 × 165	1 Rabīʿ II 1284 [ca. 2 August 1867]	*FF. PALAZZUOLI* under *Tre lune*	7 laid lines per cm. (V) and chain lines spaced 28–30 mm. apart (H)

Tre lune	Shape of Period	Orientation of zz	Copyist / Provenance	Citations	Notes
~50 mm. long, perpendicular to chains	circle	backwards	"Yaqub b. Faraj" "Shechem"	no. 87, 121–122 in Crown	ff. 73, 83, 252, 256, 261; beige in color, burnished
~73 mm. long, perpendicular to chains (cf. f. 198)	triangle	zz	"Yaqub b. Faraj" "Shechem"	no. 87, 121–122 in Crown	Near end of codex (cf. ff. 198, 260, 262, 270, 277, etc.), not as well burnished as above type
~80 mm. long, perpendicular to chains, above and only slightly to the left of name	circle	zz	يعقوب ابن هارون "Nablus"	no. 113, 149–150 in Crown	ff. 3, 7, 16, etc.
~52 mm. long, perpendicular to chains, above and to the left of name	circle	backwards (cf. ff. 12, 36)	يعقوب ابن هارون "Nablus"	no. 113, 149–150 in Crown	ff. 2, 18, 36, etc.
~53 mm. long, perpendicular to chains, above but to the left, not directly above	circle	backwards (cf. f. 7)	"Abraham b. Marhib" "Shechem"	no. 112, 148–149 in Crown	ff. 1, 5, 7; only lightly burnished
~55 mm. long, perpendicular to chains (cf. f. 7)	circle	backwards	"Benjamin b. Jacob the Danfi" "Nablus"	no. 79, 109–110 in Crown	
48–50 mm. long, perpendicular to chains (cf. f. 63)	circle	zz	يعقوب ابن هارون "Nablus"	no. 44, 61–63 in Crown	Only a few folios of this dated section (cf. ff. 62, 63, 73). Bulk of codex on two other papers.

TABLE 2 *FF. PALAZZUOLI watermarked papers* (cont.)

Shelfmark	Folio Dimensions (mm.)	Date	Watermark	Laid and Chain Line Measurements
British Library Or. 10875	218 × 154	undated	*FF. PALAZZUOLI* under *Tre lune*	6 laid lines per cm. (V) and chain lines spaced 28 mm. apart (H)
John Rylands sam. 12	213 × 160	29 Rabīʿ II 1277 [ca. 14 Nov. 1860] to 10 al-Ǧumādā I 1277 [ca. 24 Nov. 1860]	*FF. PALAZZUOLI*	(unreported)
Private collection (Jennifer Barry)	305 × 220	undated	*FF. PALAZZUOLI* under *Tre lune*	7 laid lines per cm. (H) and chain lines spaced 26–30 mm. apart (V)
Columbia University Library X893.7 Ab9		ca. 2 Rağab 1298 [ca. 31 May 1881]	*FF. PALAZZUOLI* under *Tre lune*	6 laid lines per cm. (V) and chain lines spaced 29 mm. apart (H)
CW 2481 (Michigan State)	315 × 220	1297–1311 [1880–1894]	*FF. PALAZZUOLI* under *Tre lune*	~6–7 laid lines per cm. (V) chain lines spaced 28–30 mm. apart (H)
CW 2481 (Michigan State)	315 × 220	1297–1311 [1880–1894]	*FF. PALAZZUOLI* under *Tre lune*	~9 laid lines per cm. (V) chain lines spaced 28–31 mm. apart (H)
CW 2480 (Michigan State)	223 × 158	10 Rabīʿ I 1307 [ca. 4 Nov. 1889]	*FF. PALAZZUOLI* under *Tre lune*	7 laid lines per cm. (V) chain lines spaced 28–30 mm. apart (H)

Tre lune	Shape of Period	Orientation of zz	Copyist / Provenance	Citations	Notes
~53 mm. long, perpendicular to chains, above but to the left, not directly above	circle	backwards (cf. f. 9)	"almost certainly Jacob b. Aaron" "Nablus"	no. 27, 36–37 in Crown	
(unreported)	circle	(unreported)	"Joseph, son Israel, son Ishamael, the Danafite"	Codex XII, col.163–172 in Robertson	
~47–48 mm. long, perpendicular to chains, above but to the left, not directly above	circle	zz	not specified, likely Eastern Anatolia	n/a	cf. 491–494
~53 mm. long, perpendicular to chains, above but to the left, not directly above	circle	backwards		uncatalogued	Opening of codex
~65 mm. long perpendicular to chains	circle	zz	"Jakob son of Ahron son of Schelama the priest"	Anderson 1978, 31–38	Near middle of codex
~70 mm. long perpendicular to chains	triangle	zz	"Jakob son of Ahron son of Schelama the priest"	Anderson 1978, 31–38	Near close of codex
~77 mm. long, perpendicular to chains	circle	zz	راغب ابن يعقوب ابن اسماعيل ابن ابراهيم "of the Danafite family"	Anderson 1978, 70–73	End of codex

TABLE 2 *FF. PALAZZUOLI watermarked papers* (cont.)

Shelfmark	Folio Dimensions (mm.)	Date	Watermark	Laid and Chain Line Measurements
CW 2480 (Michigan State)	223 × 158	10 Rabīʿ I 1307 [ca. 4 Nov. 1889]	*FF. PALAZZUOLI* under *Tre lune*	9 laid lines per cm. (V) chain lines spaced 29–31 mm. apart (H)
CW 2480 (Michigan State)	223 × 158	10 Rabīʿ I 1307 [ca. 4 Nov. 1889]	*FF. PALAZZUOLI* under *Tre lune*	6 laid lines per cm. (V) chain lines spaced 29–31 mm. apart (H)
CW 2486 (Michigan State)	164 × 115	12 Ǧumādā I 1305 [ca. 26 Jan. 1888]	*FF. PALAZZUOLI* under *Tre lune*	6–7 laid lines per cm. (H) chain lines spaced ~30 mm. apart (H)
CW 2486 (Michigan State)	164 × 115	12 Ǧumādā I 1305 [ca. 26 Jan. 1888]	*FF. PALAZZUOLI* under *Tre lune*	9 laid lines per cm. (H) chain lines spaced 28+ mm. apart (H)
CW 26344 (Michigan State)	214 × 146	6 Rabīʿ II 1277 [ca. 22 Oct. 1860]	*PB* with *Tre lune*	8–9 laid lines per cm. (V) chain lines spaced 22–26 mm. apart (H)

NB: Most of the Chamberlain Warren manuscripts have not been foliated or paginated by the current holding library.

Anderson, Robert T. 1978. *Studies in Samaritan Manuscripts and Artifacts—The Chamberlain-Warren Collection*, Cambridge (MA), American Schools of Oriental Research, "American Schools of Oriental Research Monograph Series" 1.

Crown, Alan David. 1998. *A Catalogue of the Samaritan Manuscripts in the British Library*, London, British Library.

Isl. Ms. 653 v. 1, 2, 4 Online catalogue records, University of Michigan Library (Evyn Kropf)

Tre lune	Shape of Period	Orientation of *zz*	Copyist / Provenance	Citations	Notes
~70 mm. long, perpendicular to chains	triangle	*zz*	راغب ابن يعقوب ابن اسماعيل ابن ابراهيم "of the Danafite family"	Anderson 1978, 70–73	Near opening of codex
~75 mm. long, perpendicular to chains	triangle	*zz*	راغب ابن يعقوب ابن اسماعيل ابن ابراهيم "of the Danafite family"	Anderson 1978, 70–73	Opening of codex
measurement impossible	circle	backwards	"Salama son of Amram the Levitical priest"	Anderson 1978, 73–75	Near opening of codex
measurement impossible	triangle	*zz*	"Salama son of Amram the Levitical priest"	Anderson 1978, 73–75	Near opening of codex
~80 mm. long, perpendicular to chains	n/a	n/a	"Joseph son of Israel son of Ishmael son of Abraham as Sarawi al Danfi"	Anderson 1978, 79–80	Near latter part of codex

Isl. Ms. 653 v. 1 | https://mirlyn.lib.umich.edu/Record/006814133/Description#tabs
Isl. Ms. 653 v. 2 | https://mirlyn.lib.umich.edu/Record/006814134/Description#tabs
Isl. Ms. 653 v. 4 | https://mirlyn.lib.umich.edu/Record/006814136/Description#tabs
Robertson, Edward. 1938. *Catalogue of the Samaritan Manuscripts in the John Rylands Library, Manchester*, Manchester, Manchester University Press.

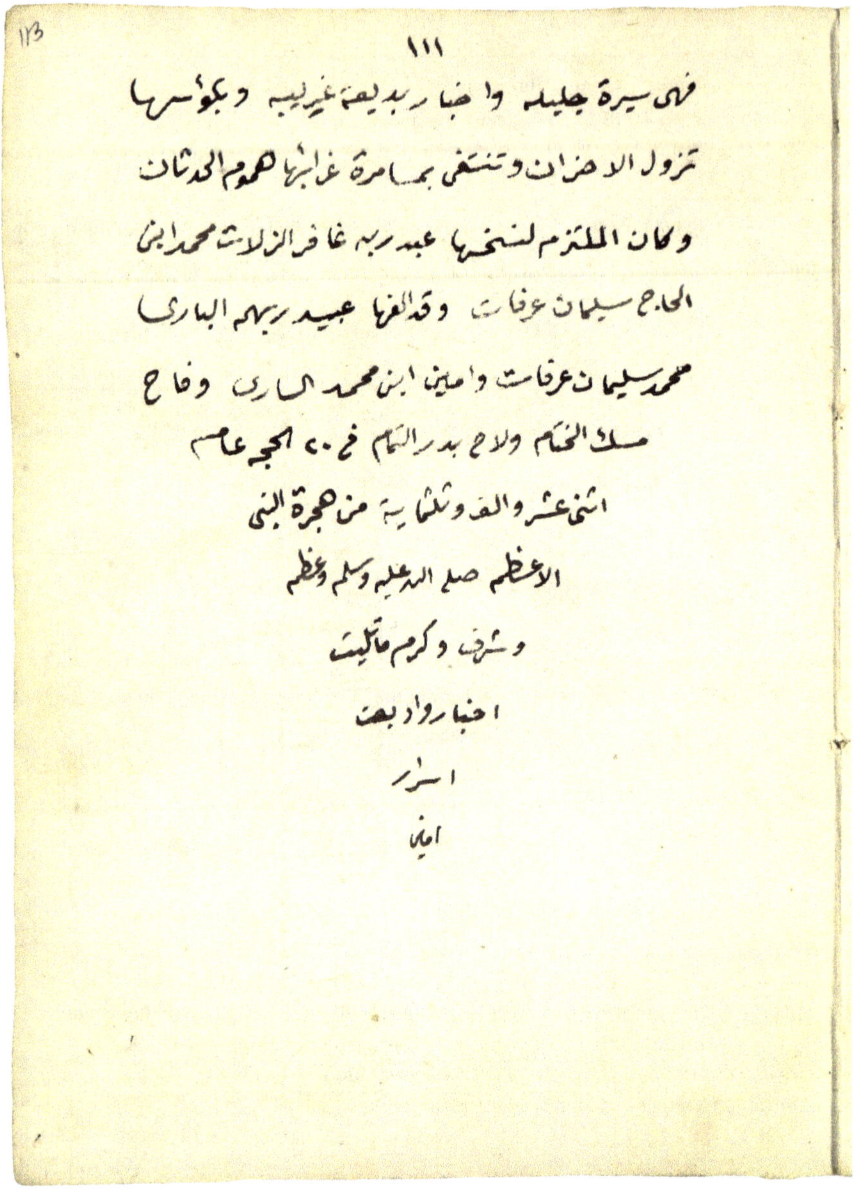
113
١١١
فهي سيرة جليلة واخبار بديعة غريبة وبكؤسها
تزول الاحزان وتنتفي بمسامرة غرائبها هموم الحدثان
وكان الملتزم لنسخها عبد ربه غافر الزلات محمد ابن
الحاج سليمان عرفات وقد الفها عبيد ربهم الباري
محمد سليمان عرفات وامين ابن محمد الباري وفاح
مسك الختام ولاح بدر التمام في ٢٠ الحجة عام
اثني عشر والف وثلثماية من هجرة النبي
الاعظم صلى الله عليه وسلم وعظم
وشرف وكرم ماكيت
اخبار واد بعث
اسرار
اميا

FIGURE 3.3 *Image of colophon in Isl. Ms. 653, vol. 24, p. 113*

If the marks indeed represent the production of Italian firms and not an elaborate form of imitation,[17] the countermark could suggest that the Alikurna designation circulated not only among the Ottoman consumers of such paper, but among 19th century Italian producers of it.

The question of manufacture for these watermarked papers is thus an important one. Establishing whether or not the Italian firms named in the watermarks are attested in documentary sources would help resolve this question, as would a fuller survey of the material evidence for similarly watermarked papers.

Investigating the material evidence of these manuscripts in light of their contextual paratexts (and vice versa) along with any documentary evidence of the location and activity these two papermaking firms could thus shed light not only on the circulation of these papers but on 19th century conceptions of Alikurna.

3 Contextual Evidence in Isl. Ms. 653

Isl. Ms. 653 is in 25 volumes and carries the text of *Sīrat Fāris al-ʿIrāq: man zalzala ǧamīʿ al-āfāq wa-šattat al-mulūk min al-ṭibāq, al-fāris al-mumāris wa-al-asad al-mutāris al-Malik Asad al-Fawāris*, a collection of the epic tales of al-Malik Asad al-Fawāris b. Ǧūdar [or Ǧawdar], the *fāris*, or mounted warrior, of ʿIrāq, in the manner of the Arabic heroic narrative (*sīra šaʿbiyya*). According to a colophon at the close of the final volume [Fig. 3.3], the collection was apparently compiled in the form attested in the manuscript by Muḥammad b. Sulaymān ʿArafāt and Amīn b. Muḥammad al-Sārī. This colophon also provides a date of completion for the transcription of the volumes—20th of Ḏū al-Ḥiǧǧa 1312/ca. 14 June 1895, and indicates that one of the compilers (Muḥammad b.

17 Vsevolod Nikolaev has attributed certain watermarked papers to Ottoman manufacture in imitation of European marks solely on the basis of their "crude, irregular, poorly designed lines and the absence of corresponding countermarks". (Cf. V. Nikolaev, *Watermarks*, 1954, 6). He later elaborates that some of these "unskillfully made" marks have in fact appeared with countermarks in Latin characters, for which he offers the wholly unsatisfactory explanation that the producer was "either of Italian origin—Venetian or Genoese, or a Spanish Jew who had emigrated from Spain, Portugal, Germany or Italy and had continued to practise the trade he had brought over from these countries, using in all probability the same counter mark, too". (Cf. ibid.). Can the paper made by such a producer truly be called Ottoman in origin? Regardless, no evidence exists for the imitation of a European name in papers of Ottoman production, particularly not in the 19th century.

Sulaymān ʿArafāt) executed the transcript. From marks appearing elsewhere in the set, we learn that he also owned the volumes.[18]

No place of copying is explicitly provided in the colophon. Certainly the hand reflects the *ruqʿa* script common across the late Ottoman lands. The name of the copyist suggests a tentative, possible link to Nablus via the well-known ʿArafāt family, active as textile merchants in the 19th century.[19] However, as acknowledged by Beshara Doumani in his study of the family, drawing upon family names as a means to trace documentary evidence is quite a challenge for this period. While the practice of omitting family names and identifying a male solely by his given name and his father's name was virtually discontinued by the end of the 19th century in the face of social and Ottoman bureaucratic forces, earlier documents reflect this practice and its ambiguity. Not only that, but family names among the middle and lower-classes, such as artisans and merchants, were rather unstable, evolving with the socioeconomic dynamics experienced by the extended family's households. In the case of the ʿArafāt family of Nablus, the four sons of al-Ḥāǧǧ ʿArafāt al-Šahīd (fl. 1720) took the name of their father as a family name and became known in the court registers as "sons of ʿArafāt". Intriguingly, rather than taking the names of their fathers as family names, the sons and grandsons of the "sons of ʿArafāt" maintained the family name "ʿArafāt", distinguishing themselves as a distinct unit bound not only by kinship but by social, economic and political ties.[20] If Muḥammad b. Sulaymān can be linked to this family, it would be as one of the sons or grandsons maintaining the ʿArafāt family name. No explicit sign of him appears in the family tree constructed by Doumani on the basis of the first fourteen volumes of the Nablus Islamic court records, though Doumani acknowledges that the ʿArafāt's official family tree differs in some respects from the record-derived version.[21]

Indeed the ʿArafāt families are known in other areas, and it is important to seek additional evidence for the manuscript's provenance. The marks of other former owners appearing in the volumes establish that the volumes circulated in Ottoman Palestine. Ownership statements in volumes 1 and 3 each reference Yāfā (Jaffa), as does the text of a letter addressed from Ṭarābulus Šām (Tripoli,

---

18 See the seal impressions appearing in vols. 1 and 6, and ownership statements (sometimes also suggesting authorship) appearing in vols. 2, 7, 8, 9, 15, and 16. Cf. Catalogue descriptions for the set and each of the vols. available in the University of Michigan Library's online catalogue Mirlyn (https://mirlyn.lib.umich.edu/).

19 Cf. "The Arafat Family", 1995, 61–68.

20 Cf. B. Doumani, *Rediscovering Palestine*, 1995, 65.

21 Cf. B. Doumani, *Rediscovering Palestine*, 1995, 279, Ch. 2, note 50, and Plate 6.

Lebanon) to one of those former owners in Yāfā which appears transcribed in volume 16.[22]

Further, these statements support the notion that the copyist could be linked to the Nablusi ʿArafāt family of textile merchants. During the latter part of the 19th century, the ʿArafāt family of Nablus also had ties to Jaffa through their trade in textiles. Doumani indicates that by this time, the trade network of Ḥāǧǧ Ismāʿīl ʿArafāt (d. 1903) had turned north toward Damascus and Beirut.[23] Relying on an account from the memoirs of Muḥammad ʿIzzat Darwaza, Doumani describes what he suggests represents a typical business trip for Nablusi textile merchants to select goods and negotiate prices in Beirut and Damascus.[24] The itinerary took them through the port city of Jaffa on their way to Beirut. This suggests that members of the ʿArafāt family might have regularly passed through Jaffa, establishing social and even commercial ties there that could have led to the exchange and sale of personal goods such as manuscripts.

4 Material Evidence in Isl. Ms. 653

Not only do the papers of Isl. Ms. 653 recall the Alikurna designation apparently circulating among the Ottomans and the firms supplying paper to them, but they help to further establish the provenance of this manuscript.

For the most part, the volumes of the manuscript have been copied on the Ferdinando Betti marked paper mentioned previously. The paper is laid with roughly 9–10 laid lines per cm. (vertical) and chain lines spaced 23–26 mm.

22 Isl. Ms. 653, vol. 1: On verso of final leaf what appears to be an ownership statement in name of al-Šayḫ Ḥasan al-Qudsī, reads "هذا الجزؤ الاول من قصة الإمير اسد الفوارس وهو من كتب الشيخ حسن القدسي في وكالة البلالية [؟] في مدينة يافة [؟]"; Isl. Ms. 653, vol. 3: Ownership statement on 'title page' (p. 3) seemingly dated 1321/1903–1904 in name of al-Šayḫ Muḥammad Uways al-Ṣāliḥ in Yāfā (Jaffa), statement reads "من كتب الشيخ محمد اويس الصالح بيافا (...) سنه ١٢٣١"; Isl. Ms. 653, vol. 16: on back flyleaf (p. 126) in pink ink, partial text of a letter to al-Šayḫ Ḥasan Afandī al-Qudsī (likely the aforementioned) in Yāfā (Jaffa) from one Ḥilmī [?] al-Ṭarābulusī in Ṭarābulus Šām (Tripoli, Lebanon).

23 B. Doumani, *Rediscovering Palestine*, 1995, 72 ff.

24 "At least once a year, Nabulsi merchants went personally to choose the types of goods they needed to stock their shops and to negotiate prices. The trip to Jaffa, Beirut, and Damascus, and back became easier and quicker with time. To Jaffa, the trip took only a few hours by carriage, compared to a day and a half on a mule before roads for wheeled transport were paved. From this ancient port city, Nabulsi merchants traveled on ships to Beirut (…)", B. Doumani, *Rediscovering Palestine*, 1995, 74.

apart (horizontal). The watermark incorporates a version of the three crescents or *Tre lune* motif so incredibly popular in the Italian papers that circulated in the Ottoman lands,[25] set above the name FERDINANDO BETTI in large capital letters in Latin script. The name is centered above the Arabic script countermark "ليكوريا" which is set in a curved flourish. [Fig. 3.4]

The *Tre lune* mark is approximately 78 mm. long, positioned perpendicular to the chains and set above the first five letters of the name FERDINANDO in the leftward portion of the mark. [See again Fig. 3.4]

The Betti firm appears to have been active in Colle di Val d'Elsa since at least 1842, at which point a "G. Betti" with one vat is reported.[26] The 1883 issue of *The paper mill directory of the world* lists a "J. Betti" of "Colle di Valle" in Siena, producing "Fine Writing" paper with "1 machine and 4 vats".[27] As for Ferdinando Betti, an 1896 directory lists "Betti Ferdinando" among the firms active in

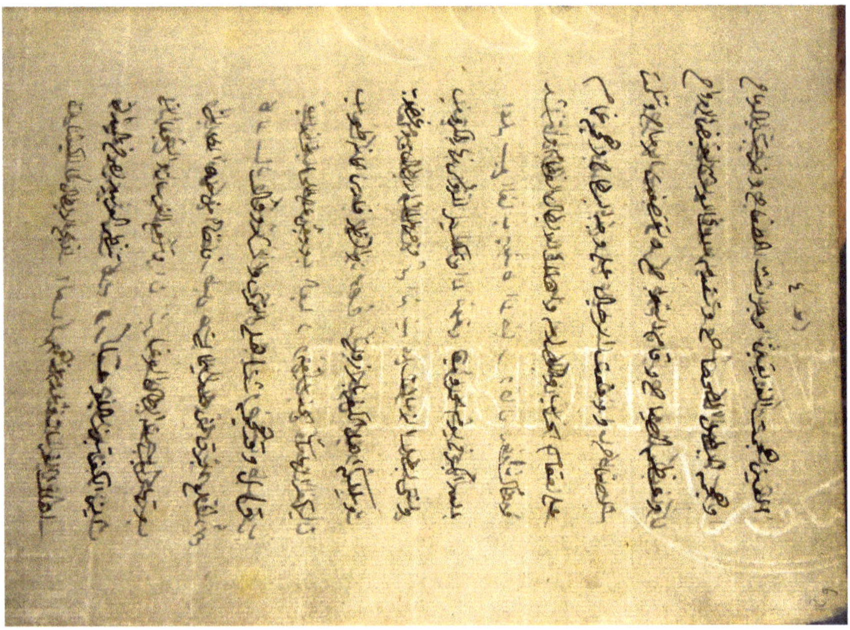

FIGURE 3.4 *Image of the leftward portion of the Betti mark appearing in Isl. Ms. 653, vol. 24, pp. 5/6 with* Tre lune *visible above the name* FERDINAN[DO] *itself above the Arabic script countermark*

25 V. Nikolaev, *Watermarks*, 1954, 4, 64, 123 ff.

26 Cf. L. Serristori, *Statistica dell'Italia*, 1842, 159.

27 Cf. *The paper mill directory of the world. A complete catalogue of all the paper and pulp mills of the globe*, 1883, 177.

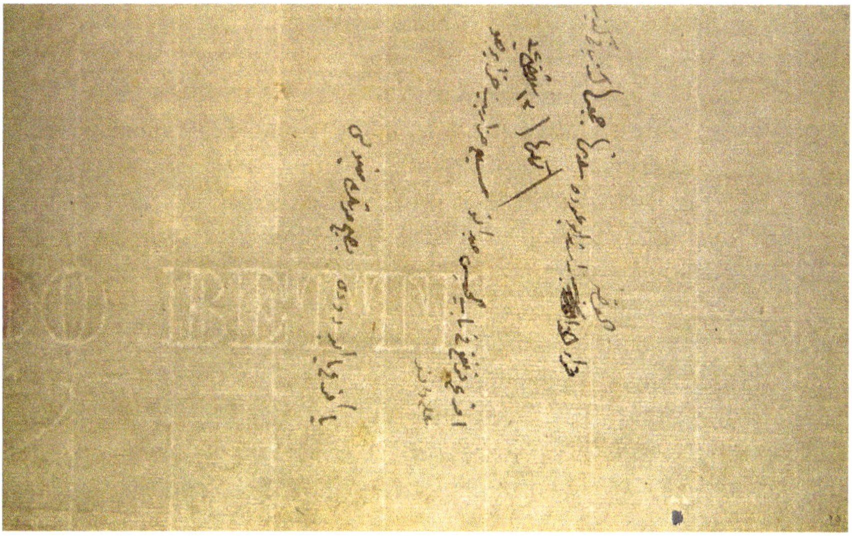

FIGURE 3.5 *Image of the rightward portion of the Betti mark appearing in Isl. Ms. 653, vol. 24, pp. 1/2 with name* [FERDINAN]*DO BETTI visible above the curved flourish in which the Arabic script countermark is set*

"Provincia di Siena (…) Colle val d'Elsa".[28] An 1898 Ministry of Agriculture statistical report also lists a "Betti Fernando" among the paper firms active in PROVINCIA DI SIENA (…) Colle Val d'Elsa.[29] Again an 1899 issue of another Italian industrial journal lists "Betti Ferdinando" in "Provincia di Siena (…) Colle Val d'Elsa".[30] Likely this is the Ferdinando Betti here being referenced in this watermark.

Thus, more likely than an attribution to Ottoman manufacture, the manufacture of this Ferdinando Betti/"ليكوريا" watermarked paper may be attributed to the Italian papermaker Ferdinando Betti of Colle di Val d'Elsa.

In volume 4, this Ferdinando Betti/"ليكوريا" watermarked paper supplements another type of paper associated with the aforementioned Fratelli Palazzuoli. This Palazzuoli marked paper exhibits about 9 laid lines per cm. (vertical) and chain lines spaced about 28–31 mm. apart (horizontal). In contrast with the Betti mark and the Palazzuoli mark observed by Ersoy, it includes no Arabic

28 Cf. "ELENCO DELLE DITTE ESERCENTI L'INDUSTRIA DELLA CARTA IN ITALIA a. Cartiere", 1898, 102–105.

29 Cf. "ELENCO DELLE DITTE ESERCENTI L'INDUSTRIA DELLA CARTA IN ITALIA a. Cartiere", 1898, 68.

30 Cf. L'INDUSTRIA DELLA CARTA IN ITALIA, *Giornale della libreria, della tipografia, e delle arti ed industrie affini*, 1899, 108.

script countermark, but also carries a watermark of the *Tre lune* variety, with the slender three crescents about 67–68 mm. long, perpendicular to chains, above and slightly to the left of the name FF. PALAZZUOLI in Latin script. The period following the FF is triangular in shape. The same Palazzuoli type exclusively comprises the text blocks of volumes 1 and 2. [Figs. 3.6–3.7]

According to the section on Tuscany in Luigi Serristori's *Statistica dell'Italia*, in 1842 the "Fratelli Palazzoli" also had a paper mill in "Colle" (presumably Colle di Val d'Elsa) with three vats.[31] Though the spelling differs from the name encountered here and in so many other 19th century watermarked papers, likely the same papermakers are intended. Reference to the name of this firm is nevertheless lacking in other industry registers.

In volume 24, the Ferdinando Betti/"ليكوريا" paper supplements still another laid paper with 12 laid lines per cm. (vertical), chain lines spaced 25 mm. apart (horizontal), and watermarks of three crescents (68 mm. long, perpendicular to chains) paired with "S S B" in a triangular shape with "B" as the point of the triangle. This firm has yet to be identified.[32]

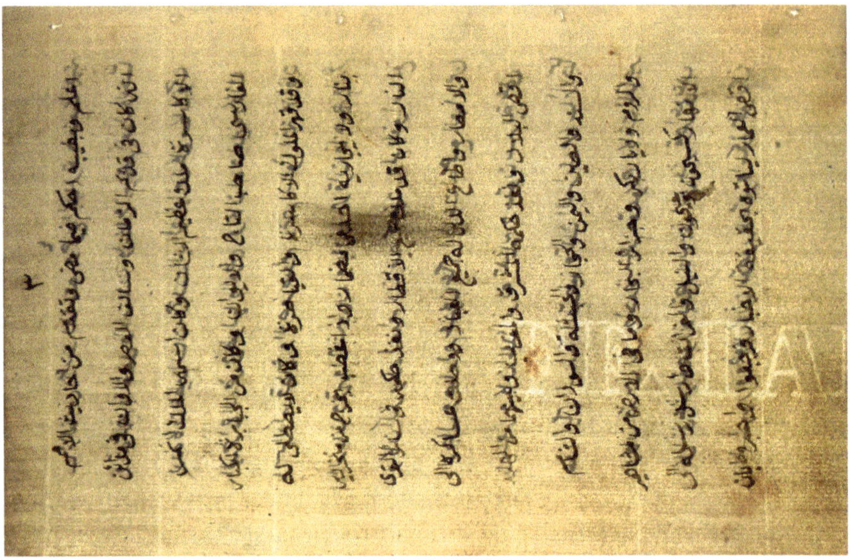

FIGURE 3.6 *View of the leftward portion of the Palazzuoli mark in Isl. Ms. 653, vol. 1, pp. 9/10, with triangular shaped period visible following the* FF

31 Cf. L. Serristori, *Statistica dell'Italia*, 1842, 159.

32 Terrence Walz apparently observed an "S S B" mark in papers of several Nigerian manuscripts, but was likewise unable to identify the responsible firm, cf. "The Paper Trade of Egypt and the Sudan", 2011, 100.

FIGURE 3.7 *View of the rightward portion of the Palazzuoli mark in Isl. Ms. 653, vol. 1, pp. 7/8*

Intriguingly, papers almost identically marked with the Betti and Palazzuoli watermarks appear in numerous Samaritan manuscripts produced in Nablus and vicinity around the same time. [Tables 1–2]

5 19th Century Circulation of Ferdinando Betti/"ليكوريا" Marked Paper

I have been able to identify at least two British Library Samaritan manuscripts that include papers with the Ferdinando Betti watermark with Arabic script countermark: British Library Or. 10723 and British Library Or. 10370. [Figs. 3.8–3.9] Manuscript Or. 10723 is undated but includes at least one bifolium of the Ferdinando Betti type among a codex comprised mainly of a Fratelli Palazzuoli type similar in morphology—that is, motifs, relative placement and measurements—to the type appearing in Isl. Ms. 653 at the University of Michigan Special Collections Library, down to the triangle-shaped period. [Table 1 and Table 2] The transcript of manuscript Or. 10370 is dated 1898 and is copied exclusively on the Ferdinando Betti/"ليكوريا" paper type. Measurements were difficult due to the small size of the manuscript and resulting positioning of the marks following folding and trimming. Nevertheless, they are quite similar to the measurements for the paper of Isl. Ms. 653 with all the familiar motifs and placement. Once again, in morphology the papers are remarkably similar.

In fact, Samaritan scribes were apparently making extensive use of this paper. As reported by Rothschild, at least eight additional Samaritan manuscripts

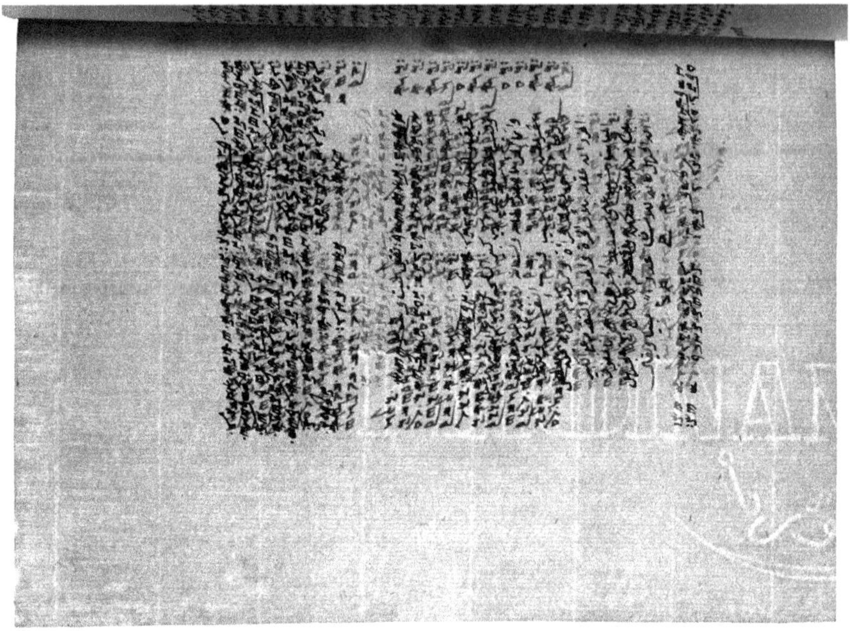

FIGURE 3.8 *View of the Ferdinando Betti/"ليكوريا" mark in folio 31r, British Library Ms. Or. 10723*
© BRITISH LIBRARY BOARD

among those held by the Bibliothèque nationale de France (BnF) also include papers with the Ferdinando Betti watermark accompanied by Arabic script countermark: BnF sam. 33, 37, 49, 50, 53, 55, 58 and 61.[33] [Table 1] Dates of transcription for these manuscripts range from 1894 to 1902.

At least one other manuscript codex from the BnF is comprised of paper with the Ferdinando Betti watermark with Arabic script countermark, BnF Ar. 6956.[34] Transcription of this multiwork codex, working from a copy in the hand of the compiler, was apparently completed 12 Šaʿbān 1312/8 February 1895 by Muḥammad Adīb b. al-Sayyid Arslān al-Taqī, who was the grandson (through his mother) of the compiler's sister. The place of transcription is apparently not specified. This manuscript was copied only a few months before

33 I have not yet been able to compare the Arabic script countermark appearing in these papers with that appearing in Isl. Ms. 653, but I suspect they are quite similar if not identical. Any variation would of course be quite interesting. Cf. J.-P. Rothschild, *Catalogue des manuscrits samaritains*, 1985, "Codicologie 1) Support", 20–21. Three other BnF Samaritan manuscripts may also include this mark, namely sam. 18, 20, and 47.

34 Cf. *Arabe 6956. Traité juridique F. 7v–36, Notice descriptive*, BnF, 2008, http://archivesetmanuscrits.bnf.fr/ark:/12148/cc8977v.

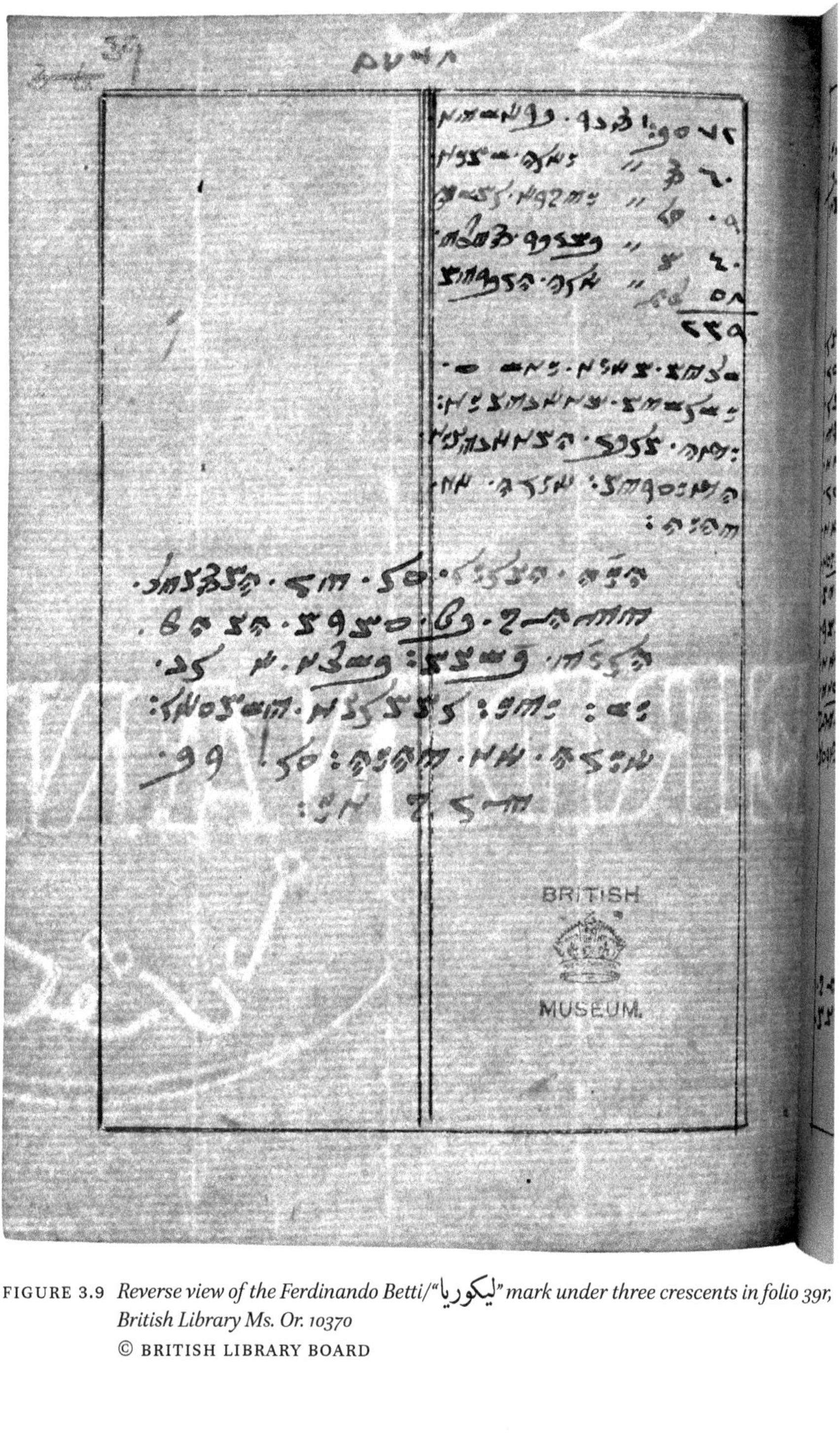

FIGURE 3.9 *Reverse view of the Ferdinando Betti/"ليكوريا" mark under three crescents in folio 39r, British Library Ms. Or. 10370*

Isl. Ms. 653. The copyist may be the father of the Damascene scholar Ḥāmid (or Muḥammad Ḥāmid) b. Adīb b. Arslān al-Taqī (d. 1967) listed by Ziriklī.[35] This could suggest a Damascene provenance for the manuscript, if the copyist was active there at the time, though certainly he could have been active elsewhere. If transcribed in Damascus, it suggests a wider circulation of the Ferdinando Betti/"ليكوريا" marked paper.

The contemporary circulation and similarity of morphology suggests that the paper being used by Muḥammad b. Sulaymān ʿArafāt was almost certainly produced by the same firm that produced the paper circulating heavily among Samaritan scribes in late 19th century Nablus.

Further, a number of the copyists employing these papers—specifically the four copyists who transcribed Or. 10723 and BnF sam. 33, 49, 50, 53, 55 58, 61—are all from the well-known Samaritan family al-Danafī.[36] Beshara Doumani claims that members of this family often served as scribes or treasurers for the rulers of Nablus and its wealthy merchants, including at least one member of the ʿArafāt family, Ḥāǧǧ Ismāʿīl ʿArafāt.[37] Portions of at least four of these manuscripts were copied by ʿAzīz b. Yaʿqūb b. Čalabī b. Yūsuf b. Čalabī al-Danafī.[38] Two of the BnF manuscripts—sam. 49 and sam. 55, dated just six months apart in 1894—were copied by Naʿmān b. Yūsuf al-Danafī. As Rothschild notes, this may be the same "Naʿman [b.] Yōsef Danafī—Servant to a Gentile" listed in the census from Rylands Samaritan MS. 328 by Robertson.[39] If so, and the age given by Robertson is correct, i.e. 30 years old in 1908, then he would have been 16 years old when transcribing the BnF manuscripts—quite young, but not implausibly so. Might he have been among the members of the Danafī family who served as scribes in the employ of the ʿArafāts? If so, might he and Muḥammad b. Sulaymān ʿArafāt drawn from the same stock of paper?

---

35 Cf. Ḫayr al-Dīn al-Ziriklī, *Al-Aʿlām*, 1980, vol. 2, 160.

36 Cf. E. Robertson, *Catalogue of the Samaritan Manuscripts in the John Rylands Library, Manchester.* Vol. 2: *The Gaster Manuscripts*, 1962, Rylands Samaritan MS. 328, column 249 and Addendum A (column 275 ff.).

37 B. Doumani, *Rediscovering Palestine*, 1995, 283, Ch. 3, note 2 "(…) the Danafi family of the Samaritan community, whose members often served as scribes or treasurers of the rulers of Nablus and its rich merchants, including Hajj Ismaʿil Arafat, whom we met in Chapter 2. Ibrahim Danafi inherited this position from his grandfather, Marjan".

38 Cf. J.-P. Rothschild, *Catalogue des manuscrits samaritains*, 1985, 159–160, "A. Index des copistes et des rédacteurs d'actes de vente", 1985; A. D. Crown, *Samaritan Scribes and Manuscripts*, 2001, 145–146; and E. Robertson, *Catalogue of the Samaritan Manuscripts*, 1962, vol. 2, Addendum A, col. 275–278.

39 Cf. J.-P. Rothschild, *Catalogue des manuscrits samaritains*, 1985, 101, and E. Robertson, *Catalogue of the Samaritan Manuscripts*, 1962, vol. 2, Addendum A, col. 275–276.

At the very least, the respective scribes may have had access to the same paper source, potentially imported through Jaffa or Beirut. The demand for imported paper in the region during the 19th century is well attested in the activity of the Samaritan scribes, and a shop selling "pens, pencils, and paper, and Birmingham wares" is attested for Nablus in the account of Mary Eliza Rogers published in 1865.[40] More likely this shop sold English paper imports rather than papers imported from Italy or even Beirut, given that it was under the supervision of the British Consul's agent, but it certainly attests to the presence of such products in the markets and the markets as a source of supply for the clearly attested demand.[41] Many of these imports would have arrived to Beirut and then reached Nablus via overland transport. While Italian and other merchant houses were active in closer ports such as Acre and Haifa, they were mainly concerned with organization of goods for export prior to the mid-1870s.[42] The port of Jaffa, even closer to Nablus, was also particularly active in the latter part of the 19th century. A marked increase in imports and exports—including both from and to Italian markets—is attested between the years 1885 and 1913.[43] The regular commercial itinerary of the 'Arafāt family merchants could have also made Italian and other imports arriving to Jaffa, Beirut or even Damascus that much more accessible. Investigation of the Nablusi archives would likely shed further light on this matter and help confirm the provenance of Isl. Ms. 653.

6 Conclusions: Recalling Alikurna

From this manuscript evidence we can see that FERDINANDO BETTI watermarked papers are attested in Samaritan manuscripts that may be firmly attributed to a provenance in the vicinity of Nablus, copied between 1894 and

40 Cf. M. E. Rogers, *Domestic Life in Palestine*, 1865, 261–262, 280–281.

41 Paper of Beiruti provenance is only tentatively attested for the following decade on the speculation of Kâğıtçı that the paper mill at Beirut mentioned in the 1886 *imtiyazname* from Sultan Abdülhamit II for the founding of the Hamidiye mill by Serkarin Osman Bey is synonymous with the firm referenced by Krawany operating 3 vats in the year 1875 in Turkey (cf. M. A. Kâğıtçı, *Kağıtçılık Tarihçesi*, 1936, 228, note 1 and F. Krawany, *Internationale Papier-Statistik*, 1910, 234). A single-machine operation of newspaper and wrapping paper manufacture is attested at Beirut for still a decade later (cf. S. C. Phillips (ed.), *The Paper Makers' Directory of all nations. Containing every Paper and Pulp Mill in the world*, 1884, 171).

42 Cf. A. Schölch, "European Penetration and the Economic Development of Palestine, 1856–82", 1982, 20.

43 Cf. G. G. Gilbar, "Economic Involvement of Palestine with the West", 1986, 191.

1902. Paper so marked is also attested in a contemporary Arabic manuscript of less certain provenance, 23 of the 25 volumes of Isl. Ms. 653, copied in 1895, possibly also in the vicinity of Nablus. Finally, the paper is attested in another contemporary Arabic manuscript of uncertain Damascene provenance, BnF Ar. 6956, also copied in 1895. All of the Ferdinando Betti watermarked papers thus far confidently attested incorporate the Arabic script countermark "ليكوريا" and the *Tre lune* motif.[44]

The watermark evidence in these and other manuscripts likewise implies that Palazzuoli watermarked paper circulated widely, from Nablus and Damascus to Siirt in Southeastern Anatolia and the vicinity of Mosul in Iraq, from at least 1860 to 1899.[45] The only confidently attested Fratelli Palazzuoli paper countermarked with "لكوريا" circulated in Southwestern Anatolia during 1871–1875.

Altogether this suggests a sophisticated production, likely beyond the capacity of contemporary Ottoman firms.[46] Only two such firms—in addition to the mill at Beirut—are attested for the period; namely, the mill in İzmir, chiefly manufacturing paper by machine, from 1843[47] and the Hamidiye Mill from 1887, again chiefly manufacturing machine-made paper.[48]

The fact that in each case the place name has been rendered in the Arabic script rather than the Latin characters used for the name of the firm suggests a deliberate attempt at recalling a designation recognizable to customers in the Ottoman lands. After all, European papermakers routinely included Latin character references to other places of manufacture in the papers they circulated there.[49]

Further, while the marks in question include Arabic script countermarks, the presence of the Latin script names of Italian firms—both of which are attested in documentary evidence to have been active in Colle di Val d'Elsa, a

44 I have also identified a watermark that may only tentatively be attributed to the Ferdinando Betti firm, a *Tre lune* mark with "FB" countermark appearing in a Samaritan manuscript of the Michigan State University Chamberlain Warren Collection, CW 2481, copied between 1880–1894. See Table 1.

45 For a more detailed discussion of this evidence for the circulation of Palazzuoli paper, again see the Appendix.

46 Osman Ersoy has offered perhaps the most thorough study to date of Ottoman papermaking activity in the 18th and 19th centuries, cf. O. Ersoy *XVIII. ve XIX. Yüzyıllarda Türkiye'de Kâğıt*, 1963.

47 O. Ersoy, *XVIII. ve XIX. Yüzyıllarda Türkiye'de Kâğıt*, 1963, 48–51.

48 O. Ersoy, *XVIII. ve XIX. Yüzyıllarda Türkiye'de Kâğıt*, 1963, 51–53; M. A. Kâğıtçı, *Kağıtçılık Tarihçesi*, 1936, 227 ff.

49 Nikolaev lists several (cf. *Watermarks*, 1954, 4, 124, 470), as does Velkov (cf. A. Velkov, *Les Filigranes dans les documents Ottomans : divers types d'images*, 2005).

town in the province of Siena in the region of Tuscany—favors Italian rather than Ottoman manufacture.

The Arabic script countermarks' crude calligraphy in a single line rather than fuller outline of the letterforms yet again suggests design and manufacture by a firm beyond the Ottoman realm, where the art of calligraphy flourished. Even in the difficult medium of the wire mould, Ottoman papermakers managed to produce calligraphy of refinement, mainly in outline rather than single line.[50] [Fig. 3.10]

Certainly the material evidence of Isl. Ms. 653 and more powerfully the material evidence of numerous Samaritan manuscripts reveals a close overlap in the circulation of the Palazzuoli and Ferdinando Betti watermarked papers. Papers bearing the marks of these firms appear together in numerous Samaritan manuscripts from the vicinity of Nablus and even Arabic manuscripts potentially of Nablusi or Damascene provenance.

Documentary evidence attesting to a shared history of operation in the same town of Colle di Val d'Elsa suggests the strong potential for borrowings or other design influence between the two firms, not to mention knowledge of the customer base. Indeed, the respective marks bear a remarkable resemblance with the name of each firm appearing below a slender *Tre lune* mark.[51] Nevertheless, the similarity and yet distinctive pointing in the case of the two Arabic script countermarks in the papers suggests somewhat independent reference to a place name or a designation, rather than blind imitation by the Ferdinando Betti firm of the earlier Palazzuoli mark.

In fact, the added pointing may suggest an attempt to improve the legibility, resonance and strength of the branding. However, the choice of a phonetic reading of Liguria—likely the coastal region of Northwestern Italy though a reference to the Ligurian Sea might also be possible—for the Arabic script countermark is somewhat surprising given the firm's location in Colle di Val d'Elsa. If geographic proximity were intended, the reading for Leghorn or Livorno allowed by the unpointed designation provided in the Fratelli Palazzuoli mark would seem far more sensible.

It may be then that the intention was indeed to recall the Alikurna paper designation known to the 17th century Ottomans and place names apparently known to the 18th century and even 19th century Ottomans.

50 See footnote 6 for a listing of references to examples of such marks.

51 Intriguingly, additional watermarks exist—only tentatively attributable to these firms—that may further suggest borrowing or imitation, namely the *Tre lune* marks with initial countermarks "BP" and "FB" appearing in Samaritan manuscripts of the Chamberlain Warren Collection of Michigan State University. See Table 1 and Table 2.

FIGURE 3.10 *Watermark 944*
REPRODUCED IN V. NIKOLAEV, *WATERMARKS OF THE MEDIAEVAL OTTOMAN DOCUMENTS IN BULGARIAN LIBRARIES*, 1954

If such a designation was intended, the discrepancy in spelling—with a preference for the more phonetic form of the place name "لكوريا" or "ليكوريا" over the vaguely derived form of the name used in Ottoman documents "عليقورنه", even as recently as 28 years prior to the date of the register in which this mark appears—would suggest design and manufacture by a firm less familiar with the language and documentary heritage.

Although a more substantial investigation of the Ottoman archives will have to be conducted in order to better establish the history of this designation, the watermark evidence presented here suggests an extraordinary attempt at branding through use of a designation specially tailored to the target market on the part of two 19th century Italian papermakers.

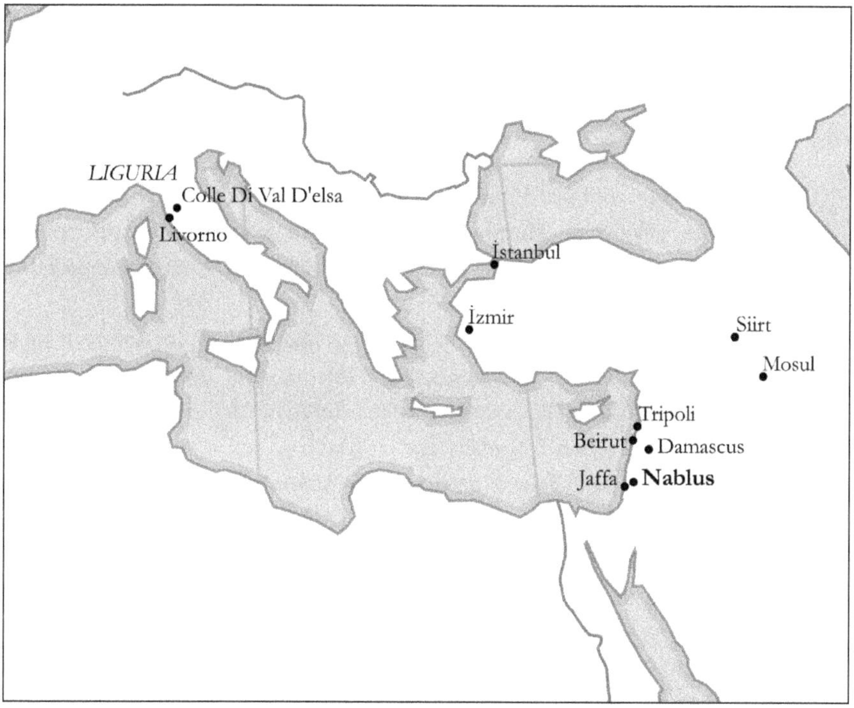

MAP 3.1 *Locations significant for the production and circulation of 19th century* "ليكوريا" *countermarked paper*
MAP PREPARED BY THE AUTHOR

Bibliography

Anderson, Robert T. 1978. *Studies in Samaritan Manuscripts and Artifacts—The Chamberlain-Warren Collection*, Cambridge (MA), American Schools of Oriental Research, "American Schools of Oriental Research Monograph Series" 1.

Babinger, Franz. 1931. *Zur Geschichte der Papiererzeugung im Osmanischen Reiche*, Berlin, Reichsdruckerei.

Babinger, Franz. 1931. APPUNTI SULLE CARTIERE E SULL'IMPORTAZIONE DI CARTA NELL'IMPERO OTTOMANO SPECIALMENTE DA VENEZIA, *Oriente Moderno* 11/8, 406–415.

Babinger, Franz. 1931. "Papierhandel und Papierbereitung in der Levante", *Wochenblatt für Papierfabrikation* 62/52, 1215–1219.

Baethgen, Friedrich. 1885. "Der Psalmencommentar des Theodor von Mopsuestia in syrischer Bearbeitung", *Zeitschrift für die Alttestamentliche Wissenschaft* 5, 53–101.

Biadi, Luigi. 1971. *Storia della città di Colle in Val d'Elsa*, Rome, Multigrafica Editrice [reprint of Firenze, 1859 ed.].

Catalogue des manuscrits arabes. 1972–. Paris, Bibliothèque nationale de France, 1972–1995, partie 2, vol. 2–4.

Crown, Alan D. 1986. "Studies in Samaritan Scribal Practices and Manuscript History: IV. An Index of Scribes Witnesses Owners and Others Mentioned in Samaritan Manuscripts with a Key to the Principal Families Therein", *Bulletin of the John Rylands University Library of Manchester* 68/2, 317–372.

Crown, Alan D. 1989. "The Morphology of Paper in Samaritan Manuscripts: A Diachronic Profile", *Bulletin of the John Rylands University Library of Manchester* 71/1, 71–93.

Crown, Alan D. 1991. "A Profile of Paper in Samaritan Manuscripts", in: Abraham Tal & Moshe Florentin (eds.), *Proceedings of the first International Congress of the Société d'études samaritaines: Tel-Aviv, April 11–13, 1988*, Tel-Aviv, Chaim Rosenberg School for Jewish Studies, Tel-Aviv University, 205–224.

Crown, Alan D. 1998. *A Catalogue of the Samaritan Manuscripts in the British Library*, London, British Library.

Crown, Alan D. 2001. *Samaritan Scribes and Manuscripts*, Tübingen, Mohr Siebeck.

Doumani, Beshara. 1995. *Rediscovering Palestine: Merchants and Peasants in Jabal Nablus, 1700–1900*, Berkeley, University of California Press.

"ELENCO DELLE DITTE ESERCENTI L'INDUSTRIA DELLA CARTA IN ITALIA a. Cartiere". 1898. Ministero di agricoltura, industria e commercio, Direzione di statistica, *Annali di statistica S.IV. No.91, Statistica Industriale. Fascicolo LXIII. Industria della Carta*, Rome, Tipografia Nazionale di G. Bertero via Umbria, 62–70.

Ersoy, Osman. 1963. *XVIII. ve XIX. Yüzyıllarda Türkiye'de Kâğıt*, [Ankara], Ankara Üniversitesi Basimevi.

Erişen-Yazıcı, Gülgün. 2001–. ALİKURNA, in: Mustafa İsen (ed.), *Türk Dünyası Edebiyat Kavramları ve Terimleri Ansiklopedik Sözlüğü*, Maltepe, Ankara, Atatürk Kültür Merkezi Başkanlığı, vol. 1, 124.

Ferrari, Dante. 1999. *Le carte della carta*, Milan, Libri Scheiwiller.

Gilbar, Gad G. 1986. "Economic Involvement of Palestine with the West", in: David Kushner (ed.), *Palestine in the Late Ottoman Period: Political, Social and Economic Transformation*, Jerusalem, Yad Izhak Ben-Zvi.

Graham-Brown, Sarah. 1982. "The Political Economy of the Jabal Nablus, 1920–1948", in: Roger Owen (ed.), *Studies in the Economic and Social History of Palestine in the Nineteenth and Twentieth Centuries*, Oxford, St Antony's, 88–176.

Güleç, İsmail. 1994. "Osmanlılarda kâğıt ve kağıtçılık", *Müteferrika* 2, 85–94.

Hershlag, Zuri Yehuda. 1980. *Introduction to the Modern Economic History of the Middle East*, Leiden, E. J. Brill.

L'INDUSTRIA DELLA CARTA IN ITALIA. 1899. *Giornale della libreria, della tipografia, e delle arti ed industrie affini* 12, 107–108.

Issawi, Charles (ed.). 1966. *The Economic History of the Middle East 1800–1914*, Chicago, University of Chicago Press.

Kâğıtçı, Mehmet Ali. 1936. *Kağıtçılık Tarihçesi*, Istanbul, Kader Basımevi.

Kâğıtçı, Mehmet Ali. 1976. *Historical Study of Paper Industry in Turkey/Historique de l'industrie papetière en Turquie*, Istanbul, Grafik Sanatlar Matbaası.

Krawany, Franz. 1910. *Internationale Papier-Statistik*, Berlin, Verlag für Fachliteratur.

Kurzes Verzeichniss der Sachau'schen Sammlung syrischer Handschriften. 1885. Berlin, A. W. Schade's Buchdruckerei.

Mar(I) Eliâ of SôbHâ. 1887. *A Treatise on Syriac Grammar by Mar(I) Eliâ of SôbHâ*. Richard J. H. Gottheil (ed.), Berlin, Wolf Peiser Verlag.

Niessen, Friedrich. 2010. "Samaritans Under Muslim Rule", in: Norman A. Stillman et al. (ed.), *Encyclopedia of Jews in the Islamic World*, Leiden, E. J. Brill, vol. 4, 232–235.

Nikolaev, Vsevolod. 1954. *Watermarks of the Mediaeval Ottoman Documents in Bulgarian Libraries. Watermarks of the Ottoman Empire*, Sofia, Bulgarian Academy of Sciences, vol. 1 [original title in Bulgarian language].

Nimr, Iḥsān. 1937. *Tārīkh Ǧabal Nābulus wa-al-Balqā'. Al-Juz' al-awwal: Ḥawādith ʿahd al-iqṭāʿ*, Damascus, Maṭbaʿat Ibn Zaydūn.

Özen, Mine Esiner. 1985. *Yazma kitap sanatları sözlüğü*, Istanbul, İstanbul Üniversitesi, Fen Fakültesi.

Özen, Mine Esiner. 2003. "Kağıt filigranları ve Dr Emel Esin Kütüphanesi Yazmalarından Örnekler", *Festschrift in Honor of Günay Kut, Journal of Turkish Studies* 27/II, 133–152.

The Paper Mill Directory of the World. A Complete Catalogue of All the Paper and Pulp Mills of the Globe. 1883. Holyoke (MA), C. W. Bryan & Co.

Phillips, S. C. (ed.). 1884. *The Paper Makers' Directory of all nations. Containing every Paper and Pulp Mill in the world*, London, "Paper Makers' Circular Office".

Quataert, Donald. 1983. *Social Disintegration and Popular Resistance in the Ottoman Empire, 1881–1908: Reactions to European Economic Penetration*, New York, New York University Press.

Quataert, Donald. 2005. *The Ottoman Empire 1700–1922*, Cambridge, Cambridge University Press [2nd ed.].

Rafeq, Abdul-Karim. 1973. "Les registres des Tribunaux de Damas comme source pour l'histoire de la Syrie", *Bulletin d'études orientales* 26, 219–226.

Ramīnī, Akram. 1979. *Nābulus fī al-qarn al-tāsiʿ ʿašar: dirāsa mustaẖlaṣa min siğillāt al-Maḥkama al-Šarʿīya bi-Nābulus*, Amman.

Refik, Ahmet. 1930. *Hicrî On İkinci Asırda İstanbul Hayatı (1100–1200)*, İstanbul, Devlet Matbaası.

Refik, Ahmet. 1988. *Istanbul hayatı*, Istanbul, Enderun yayınları, vol. 1–4 [reprint].

Regourd, Anne. 2014. "Introduction to the Codicology: Papers, Ruling, and Bindings", in: Alessandro Gori & coll., *A Handlist of the Manuscripts in the Institute of Ethiopian Studies*. Vol. 2: *The Arabic Materials of the Ethiopian Islamic Tradition*, Eugene (OR), Pickwick Publications, « Ethiopian Manuscrits, Texts, and Studies Series » 20, xlvii–xcii.

Robertson, Edward. 1938. *Catalogue of the Samaritan Manuscripts in the John Rylands Library, Manchester*, Manchester, Manchester University Press, vol. 1.

Robertson, Edward. 1962. *Catalogue of the Samaritan Manuscripts in the John Rylands Library, Manchester*. Vol. 2: *The Gaster Manuscripts*, Manchester, Manchester University Press.

Rogers, Mary Eliza. 1865. *Domestic Life in Palestine*, Cincinnati, Poe & Hitchcock.

Rothschild, Jean-Pierre. 1985. *Catalogue des manuscrits samaritains*, Paris, Bibliothèque nationale, Département des manuscrits.

Sabbatini, Renzo. 1990. *Di bianco lin candida prole: La manifattura della carta in eta moderna e il caso toscano*, Milan, Franco Angeli Libri.

Sayar, Ahmet Güner. 1988–. MUALLİM CEVDET, *Türkiye Diyanet Vakfı İslâm ansiklopedisi*, Istanbul, Türkiye Diyanet Vakfı, İslâm Ansiklopedisi Genel Müdürlüğü, vol. 30, 314.

Schwarb, Gregor. THE GENIZAH: AN UNDEREXPLOITED SOURCE FOR THE INTELLECTUAL HISTORY OF THE MEDIEVAL WORLD OF ISLAM (NEW FRONTIERS IN ISLAMIC STUDIES: German-Israeli Summer School funded by the Einstein-Stiftung, Berlin, 10–15 September 2012). Session III: Samaritan Manuscripts", 10.09.2012, available online at: http://www.academia.edu/1932885/Samaritan_Genizot_and_Samaritan_Manuscripts.

Schölch, Alexander. 1982. "European Penetration and the Economic Development of Palestine, 1856–82", in: Roger Owen (ed.), *Studies in the Economic and Social History of Palestine in the Nineteenth and Twentieth Centuries*, Oxford, St Antony's, 10–87.

Serristori, Luigi. 1842. *Statistica dell'Italia*, Firenze, Stamperia Granducale.

Shunnar, Zuhair. 1974. *Katalog samaritanischer Handschriften*. Vol. 1: *British Museum (Gaster Collection). Bibliothek des samaritanischen Hohenpriesters, Nablus. Orient-Institut, Leningrad. Seminar für Semitistik u. Arabistik der Freien Universität, Berlin. Library of Congress, Washington, D.C. Yale University Library, N.H. New York Public Library, N.Y.*, Berlin-West, R. Seitz.

Sprengling, Martin. 1916. "Antonius Rhetor on Versification with an Introduction and Two Appendices", *The American Journal of Semitic Languages and Literatures* 32/3, 145–216.

Tekin, Şinasi. 1993. *Eski Türklerde Yazı, Kâğıt, Kitap ve Kâğıt Damgaları*, Istanbul, Eren.

Ünver, A. Süheyl. 1962. "XV. Yüzyılda Türkiye'de Kullanılan Kâğıtlar ve Su Damgaları", *Belleten* 26/104, 739–762.

Velkov, Asparouh. 2005. *Les Filigranes dans les documents Ottomans: divers types d'images*, Sofia, Éditions "Texte—A. Trayanov".

Velkov, Asparouh & Stephane Andreev. 1983. *Filigranes dans les documents ottomans*. Vol. 1: *Trois croissants*/Водни знаци в Османотурските документи. I Три луни, Sofia, Bibliothèque nationale "Cyrille et Methode", Commission archéographique bulgare Section orientale.

Vil'sker, Leĭb Khaimovich & Lebedev, Viktor Vladimirovich (i.e. Вильскер, Лейб Хаимович и Лебедев, Виктор Владимирович). 1992. Самаритянские документы Государственной публичной библиотеки имени М.Е. Салтыкова-Щедрина. Каталог, Saint Petersburg, Государственная Публичная библиотека им. М. Е. Салтыкова-Щедрина.

Walz, Terence. 2011. "The Paper Trade of Egypt and the Sudan in the Eighteenth and Nineteenth Centuries and its Re-export to the Bilad as-Sudan", in: Graziano Krätli & Ghislaine Lydon (eds.), *The Trans-Saharan Book Trade: Manuscript Culture, Arabic Literacy, and Intellectual History in Muslim Africa*, Leiden, E. J. Brill, 73–108.

White, Charles. 1846. *Three Years in Constantinople; or, Domestic Manners of the Turks in 1844*, London, Henry Colburn [2nd ed.].

Al-Ziriklī, Ḫayr al-Dīn. 1980. *Al-Aʿlām, qāmūs tarāǧim li-ašhar al-riǧāl wa-al-nisāʾ min al-ʿArab wa-al-mustaʿribīn wa-al-mustašriqīn*, Beirut, Dār al-ʿilm li-al-malāyīn.

Appendix: 19th Century Circulation of Watermarked Paper of the Fratelli Palazzuoli

As we have already seen reported by Osman Ersoy, a mark bearing the name of the FRATELLI PALAZZUOLI in Latin characters with "لكوريا" in Arabic script below appears in the court registers of Siirt in Southeastern Anatolia, dated 1871–1875.[52] Another potential instance of this mark has been observed in a Syriac manuscript from the Semitic Museum of Harvard University, namely Semitic Museum no. 4059, apparently transcribed in Mosul in 1895.[53] According to Martin Sprengling, "The heavy paper, of a kind much in vogue in the modern Levant, bears the watermark of the Fratelli Palazzuoli in Latin and Arabic characters".[54] While this recalls the mark depicted by Ersoy, it could also be the case that multiple paper types make up the codex, some bearing the Fratelli Palazzuoli mark, others the Arabic characters, as has been observed in Isl. Ms. 653 and several other manuscripts.[55]

52 Cf. O. Ersoy, *XVIII. ve XIX. Yüzyıllarda Türkiye'de Kâğıt*, 1963, 21, note 5, 56, insert between 68–69, and 194.

53 Cf. M. Sprengling, "Antonius Rhetor on Versification with an Introduction and Two Appendices", 1916, 204, Appendix II.

54 M. Sprengling, "Antonius Rhetor on Versification with an Introduction and Two Appendices", 1916, 203.

55 Given the date of the transcript, it seems likely that the Palazzuoli paper could be accompanied by Ferdinando Betti/"ليكوريا" watermarked paper. Certainly examination of the codex could clarify the situation, which unfortunately I have not yet been able to do.

As I have observed in the three volumes of Isl. Ms. 653—copied in 1895, possibly in Nablus—a more commonly encountered mark of the Fratelli Palazzuoli involves the abbreviated form of the name FF. PALAZZUOLI again in large Latin characters, under a *Tre lune* mark of slender crescents. Versions of this mark have also been observed in two Syriac manuscripts, both copied in 'Tel Kepe in 1882.[56] Circulation of FF. PALAZZUOLI marked paper is also attested for late 19th century Damascus.[57]

Further, as is the case with the Ferdinando Betti/"ليكوريا" paper of Isl. Ms. 653, versions of FF. PALAZZUOLI paper also appear in 19th century Samaritan manuscripts. In fact, paper carrying this name is quite common among these manuscripts.[58] Among the earliest manuscripts carrying a version of the FF. PALAZZUOLI mark is John Rylands Library sam. 12, transcribed by Yūsuf b. Isrā'īl b. Ismā'īl al-Sāmirī al-Danafī in 1860.[59] Another early appearance of the mark is in British Library Or. 12257, in a section dated 1867.[60] Rothschild has identified FF. PALAZZUOLI watermarked paper in at least 10 Samaritan manuscripts at the Bibliothèque nationale, including one manuscript copied by the priest Ya'qūb b. Hārūn in 1884, BnF sam. 23.[61] These manuscripts range in date from 1866 to 1899. Still other instances of this version of the abbreviated mark appear in Samaritan manuscripts at the British Library and Michigan State University Chamberlain Warren Collection.[62] [Table 2] All are from the final two decades of the 19th century.

This preponderance of Palazzuoli marked paper in Samaritan manuscripts reveals that there exist at least three different versions of the FF. PALAZZUOLI mark. [Table 3]

---

56 Berlin Mss Sachau 215 and 216, cf. *Kurzes Verzeichniss der Sachau'schen Sammlung syrischer Handschriften*, 1885, 20; F. Baethgen, "Der Psalmencommentar des Theodor von Mopsuestia in syrischer Bearbeitung", 1885, 53–54, and *A Treatise on Syriac Grammar by Mar(I) Eliâ of Sôb^{H}â*, R. J. H. Gottheil (ed.), 1887, 20.

57 Cf. A. Rafeq, "Les registres des Tribunaux de Damas comme source pour l'histoire de la Syrie", 1973, 221.

58 Crown claims that "Palazzuoli" is among the most commonly encountered names appearing in the watermarked paper of 19th century Samaritan manuscripts, cf. A. D. Crown, *Samaritan Scribes and Manuscripts*, 2001, 106.

59 Cf. E. Robertson, *Catalogue*, 1962, vol. 1, Codex XII, col. 163–172.

60 Crown also reports this instance but without clarifying in which section the mark appears. Cf. A. D. Crown, *A Catalogue of the Samaritan Manuscripts in the British Library*, 1998, 61–63, no. 44.

61 Cf. J.-P. Rothschild, *Catalogue des manuscrits samaritains*, 1985, 96.

62 Described in R. T. Anderson, *Studies in Samaritan Manuscripts and Artifacts—The Chamberlain-Warren Collection*, 1978.

In one version, the period following the letters FF is provided in the form of an open circle and the letters ZZ appear in the name backward. [Fig. 3.11] This version of the mark appears in the first half of the 1880s and in the 1870s as well. A contemporary version with the period in the form of a circle and letters ZZ properly oriented is also attested. [Fig. 3.12]

TABLE 3 *Watermarks of the Fratelli Palazzuoli*

Watermark	Attested Dates	Attested Places of Circulation
Tre lune with		
PB countermark	1860	Nablus
FRATELLI PALAZZUOLI	1871–1875	Siirt
"لكوربا"	(1895?)	(Mosul?)
FF○ PALA ƸƸUOLI	1872, 1880, 1883,	Nablus
with Tre lune	1885–1886, 1888	
FF○ PALAZZUOLI	1883, 1889, 1894	Nablus
with *Tre lune*		
FF△ PALAZZUOLI	1885–1886, 1888, 1889,	Nablus
with *Tre lune*	1894, 1895	(Nablus or Damascus?)

FIGURE 3.11
FF. PALAZZUOLI mark with ZZ backward appearing in the paper of a Samaritan manuscript dated 2 Rağab 1298 [ca. 31 May 1881]
X893.7 AB9, RARE BOOK & MANUSCRIPT LIBRARY, COLUMBIA UNIVERSITY IN THE CITY OF NEW YORK

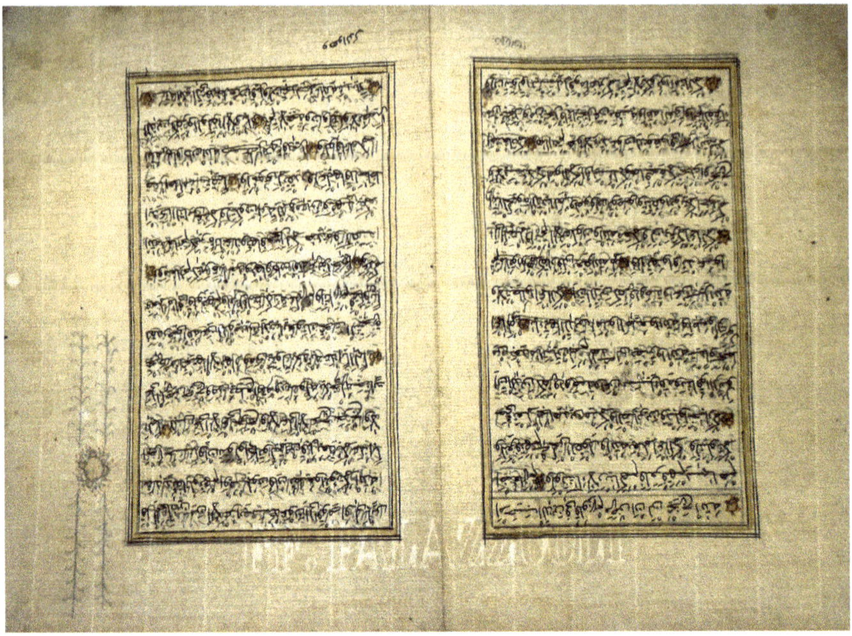

FIGURE 3.12 FF. PALAZZUOLI *mark in an undated copy of the Qur'ān likely produced in Eastern or Southeastern Anatolia*
PRIVATE COLLECTION OF JENNIFER BARRY, 491–494

In another version, the period following the FF is provided in the form of an open triangle and the letters ZZ of the name are properly oriented. This version of the mark appears in manuscripts from the second half of the 1800s and 1890s as well.

I have identified yet another mark that may be attributable to the Fratelli Palazzuoli. This mark appears in a few leaves of Michigan State University CW 26344 transcribed in 1860.[63] It consists of a *Tre lune* mark roughly 80 mm. long, perpendicular to the chain lines with the initials "PB" below and slightly to the left. Here "PB" may stand for "Palazzuoli Brothers". A countermark of similar abbreviation may be attributable to the Ferdinando Betti firm, that is "FB" (for "Ferdinando Betti") below and to the right of a *Tre lune* mark.[64]

63 Cf. R. T. Anderson, *The Chamberlain-Warren Collection*, 1978, 79–80.

64 I have observed this mark in yet another Samaritan manuscript of the Michigan State University Chamberlain Warren Collection, CW 2481, copied between 1880 and 1894 (cf. R. T. Anderson, *The Chamberlain-Warren Collection*, 1978, 31–38).

CHAPTER 4

Manuscrits de la mer Rouge (première moitié du XXe siècle) : papiers Abū Šubbāk du Yémen et d'Éthiopie[1]

Anne Regourd

Abstract

This is a statistically-based study that seeks to identify the period during which the various papers Abū Šubbāk were used in Yemen and Ethiopia. They have been classified according to a typology based on the description of internal marks of the paper. It shows that, in the first half of the twentieth century, these papers were far from being 'exotic' but were commonplace, and were a well-known product in the countries of the Red Sea where the copying of manuscripts was still carried on. Several papers shared the market, and some of the manufacturers, among them the famous Italian firm of Galvani, were Western. The area was served by more than one entry point, but the study focuses on the indirect and above all internal routes of the papers to which manuscripts offer privileged access.

1 Cette étude n'aurait pu avoir lieu sans le soutien du Centre français d'Archéologie et de Sciences sociales (CEFAS, Sanaa, Programme de sauvegarde des manuscrits des bibliothèques de Zabid sous ma dir. scientifique depuis 2001, désormais noté PZ), de la Fondation Hā'il Saʿīd et de la bienveillance de son Directeur, Fayṣal Saʿīd Fārih (Taez, mission d'octobre 2010), du Centre français d'études éthiopiennes (CFEE) et de son Directeur, Éloi Ficquet (mission de décembre 2010), de l'Institute of Ethiopian Studies (IES, Addis Ababa, mission de décembre 2010), du Sherif Museum, Harar City, Abdullah Sherif (décembre 2010), de l'UMR 7192 du Cnrs (analyse de papiers en laboratoire, novembre 2011), de la Fondation Furqān, du Ethiopic Manuscripts Imaging Project (EMIP) et de son Directeur Steve Delamarter (George Fox University, février 2012 ; mission de mai 2014), du Projet European Research Council (ERC) « Islam in the Horn of Africa. A Comparative Literary Approach », Advanced Grant no. 322849, Université de Copenhague, et de son Directeur Alessandro Gori (mission de mars-avril 2015). Une bourse Brill (« Brill Fellowship ») m'a permis de mener des recherches sur les papiers Abū Šubbāk dans les manuscrits des collections de la Bibliothèque de l'Université de Leyde durant deux mois, en juin et en septembre 2015. Les papiers dont il est question ici ont fait l'objet de plusieurs communications préliminaires, dès 2011.

 | DOI 10.1163/9789004360877_006

Introduction

Les papiers objets de l'enquête sont ceux qui portent en marque interne « Abū Šu/ibba/āk ». Une étude préliminaire avait permis d'établir que, contrairement à ce que l'on croyait jusque-là, il y avait plusieurs papiers Abū Šubbāk (A/B à E) – et non un seul[2]. L'identification et la classification d'exemples datés, en nombre encore restreint, avait néanmoins permis de se faire une première idée de la période à laquelle ces papiers avaient été utilisés au Yémen et en Éthiopie. L'autre ambition de l'étude était de poser le problème de la circulation de ces papiers, en estimant leur point de départ – ou lieu de leur production – et en notant, le cas échéant, leur point d'arrivée. À côté de leur période d'utilisation, avait peu à peu émergé une géographie de leur consommation. Méthodologiquement, on insistait donc sur la valeur documentaire du papier (ou du papier comme source) pour contribuer à une histoire de son commerce sur les rives de la mer Rouge aux xixe et xxe s.

C'est avec une base statistique bien plus ample que l'on reprend ici la réflexion. Une nouvelle enquête s'imposait avec comme objectif d'étoffer la base statistique de manière à consolider la fourchette de datation pour l'usage des papiers Abū Šubbāk, toujours au Yémen et en Éthiopie, et à obtenir ainsi un nuage statistique pour les papiers A/B à D. Les résultats dépassent de loin cet objectif en mettant clairement en évidence l'importance de la consommation relative des papiers Abū Šubbāk.

Après avoir rappelé les caractéristiques des types A à D, seront présentés et commentés le corpus augmenté, ainsi que les résultats de cette seconde enquête.

Constitution d'un corpus

Un papier ottoman

Lorsqu'il est isolé la première fois en 1997 à Zabid (Yémen), le papier Abū Šubbāk est présenté comme un exemple de ces papiers tardifs à filigranes autres qu'italiens, « fabriqués en Turquie ottomane du début du xixe siècle au (milieu ? du) xxe siècle ». Il est cité aux côtés d'un « autre papier ottoman »,

2 Voir A. Regourd, « Paper Trade on the Red Sea (XIXe c.–first half of the XXe c.): 'local papers' against Italian's. The case of an 'Ottoman' watermarked paper used in Yemen and in Ethiopia », remis en 2012, à paraître.

à contremarque en *osmanlı* mentionnant Istanbul, interprété comme un indice de la présence de papiers ottomans au Yémen jusqu'au xxe s.[3].

En ce qui concerne la présente étude, on retiendra que, considéré sous l'angle d'une classification historique mettant en présence, d'une part, des papiers italiens produits sur la longueur et, d'autre part, le groupe des papiers turc-ottomans apparus tardivement – dont en 1997, on ne connaissait que peu d'exemples, le papier Abū Šubbāk, se rapportant à cette dernière catégorie, ait pu apparaître comme un phénomène tant soit peu marginal.

Premiers relevés

Entre 2001 et 2010, 17 exemples de papier Abū Šubbāk ont été recensés dans 16 livres et un document provenant des bibliothèques privées de Zabid, s'ajoutant aux 2 signalés en 1997[4]. Ces 16 livres manuscrits sur un total de 172 en 86 volumes (24 sont des collections), dont les colophons s'échelonnent entre le milieu du xviie et le milieu du xxe s., appartiennent à la Bibliothèque de ʿAbd al-Rahman al-Hadhrami ; les textes datés au colophon apparaissent sous la cote m/ḥ dans le tableau récapitulatif (section iii.)[5]. Très vite, l'utilisation du papier Abū Šubbāk en feuillet de garde des manuscrits a été observée[6]. Entre 2009 et 2010, parmi une centaine de papiers saisis à main levée à Dār

3 G. Humbert, « À la recherche de manuscrits dans les bibliothèques privées du Yémen : première mission à Zabid », 1997, 34 ; *id.*, « Le manuscrit arabe et ses papiers », 2002, 55–77, 67. Deux des quatorze manuscrits qu'elle a étudiés, à la Bibliothèque des Ašāʿir, à Zabid, étaient écrits sur un papier filigrané à profil lunaire, portant une contremarque, en caractère *osmanlı* : « *Beyaz Abû* (sic) *Shabbâk Stambûlî* » (« papier d'Abū Šabbāk d'Istanbul »).

4 L'année 2001 correspond au coup d'envoi du pz. G. Humbert, « À la recherche de manuscrits dans les bibliothèques privées du Yémen : première mission à Zabid », 1997, 34.

5 Voir A. Regourd, « Les routes commerciales entre Zabîd et l'Europe : les papiers filigranés de fonds manuscrits de Zabîd (Yémen, fin 18e–milieu 20e s.) », 2006, 173–198 ; *id.* (dir.), *Catalogue cumulé des bibliothèques de manuscrits de Zabid. I. Bibliothèque ʿAbd al-Rahman al-Hadhrami*, fasc. 1, 2006 ; les papiers filigranés ont été reproduits dans : *id.*, avec la collaboration d'Hélène C. David et de ʿAbd al-R. al-Ahmar, *Catalogue cumulé des bibliothèques de manuscrits de Zabid. I. Bibliothèque ʿAbd al-Rahman al-Hadhrami*, fasc. 1, *Les papiers filigranés*, 2008, mss m/ḥ 1, pl. 001–003, et m/ḥ 17, pl. 126 ; *id.* (dir.), *Catalogue cumulé des bibliothèques de manuscrits de Zabid. I. Bibliothèque ʿAbd al-Rahman al-Hadhrami*, fasc. 2, 2008 ; *id.* (dir.), *Catalogue cumulé des bibliothèques de manuscrits de Zabid. I. Bibliothèque ʿAbd al-Rahman al-Hadhrami*, fasc. 3, à paraître. Les catalogues de filigrane du fasc. 2 et 3 ne sont pas parus. Les relevés à main levée des papiers filigranés, sur lesquels s'appuie notre décompte, ont été réalisés à Zabid par ʿAbd al-Rahman al-Ahmar, membre de l'équipe du pz. Le *Catalogue cumulé des bibliothèques de manuscrits de Zabid* sera désormais cité sous son acronyme *ccbmz*, suivi du tome et du fascicule.

6 *ccbmz*, I. 1, 2006, sous la cote m/ḥ 17.

al-maḫṭūṭāt, Sanaa, figurait le papier qui nous intéresse. Cette bibliothèque publique acquiert principalement ses manuscrits par don et achat. Ils peuvent donc provenir de différentes régions du Yémen, mais il est sûr que les environs de la capitale en constituent une source régulière. Enfin en octobre 2010, 12 des 97 manuscrits examinés par mes soins sur un total de 334 possédés par la Fondation Hā'il Sa'īd à Taez, contenaient du papier Abū Šubbāk[7]. Ils portent les cotes : ms. 27, 51, 57, 70, 71 à 75, 97, 106 et 107 ; deux textes sur douze sont datés au colophon, les manuscrits 27, copié en 1073/1662–1663, et 107, copié en 1328/1910–1911, mais seul le second nous intéresse, car, dans le premier cas, la présence d'un papier Abū Šubbāk à cette date s'explique par le fait qu'il ait servi de feuillet de garde.

Le papier du ms. Ar. BnF 7084, un recueil de poésie *ḥumaynī* légué par Georges Séraphin Colin à la Bibliothèque nationale de France, a été rédigé sur du papier Abū Šubbāk[8]. Il n'est pas daté au colophon, mais la page de titre porte une marque d'achat : « Ṣan'ā' Déc. 1929 / acheté 18 Riyāl = 18 × 12 = 216 frcs / environ 800 pages (400 feuillets) »[9]. Au f. 138v, figure une date de copie, 1340/1922. Aux feuillets 133v–135r, un poème (*madaḥ*) chante les louanges des Ottomans, i. e. Mustafa Kemal et le Sultan, pour leur victoire sur les Grecs, qui prit place en 1340/1921–1922. Le poème est suivi par une controverse sur la consommation de qat (f. 135r à 138v) datée de 1340/1921–1922. Enfin, entre les f. 226 et 227, se trouve une feuille de calendrier volante, à la date du 8 rabī' al-aḫar 1330, soit 1912, et, entre les f. 328 et 329, un formulaire volant en français et en turc-ottoman pour envoi de télégramme : la date pré-imprimée qui ne mentionne pas les unités, à compléter par l'utilisateur de ce service, est : 192 …[10]

7 Les manuscrits viennent des régions environnantes, mais pas exclusivement. Le chiffre de 334 manuscrits à la date d'octobre 2010, nous a été communiqué par le bibliothécaire en chef de la Fondation, sur la base de l'inventaire électronique, tenu à jour. Pour avoir une idée de l'accroissement potentiel du fonds sur une année, l'inventaire de la bibliothèque mis en ligne en février 2009, dans les *Chroniques du manuscrit au Yémen* 7, indiquait 304 manuscrits, http://www.cefas.com.ye/spip.php?article172.

8 Voir le catalogue en ligne, http://archivesetmanuscrits.bnf.fr/ead.html?id=FRBNFEAD000089778. Un historique et un inventaire des manuscrits yéménites de la collection Colin se trouve dans : M.-G. Guesdon, « La Collection Colin à la Bibliothèque nationale de France », janvier 2013. J. Dufour, « La *safīna* de Colin : une source importante pour l'histoire de la poésie *ḥumaynī* », janvier 2013, a donné une première description détaillée du contenu et de l'intérêt de la collection de poèmes réunis dans le ms. 7084.

9 Les barres obliques ont été rajoutées pour marquer le changement de ligne.

10 Un type de formulaire que G. S. Colin a pu être amené à utiliser, lors de son court séjour au Yémen en 1929, alors qu'il était Directeur d'études à l'Institut des hautes études marocaines de Rabat.

Les dimensions d'un feuillet sont de 24,5 × 18 cm. La distance entre deux fils de chaînettes mesurée à partir des bords internes est de 2,5 cm. Vingt vergeures mesurées à partir des bords externes, couvrent 3,4 cm.

Un papier répandu au Yémen

C'est cependant un séjour d'étude à la Bibliothèque de l'Université de Leyde (Universiteitsbibliotheek Leiden, Leyde, Pays-Bas, désormais UbL), connue pour ses riches collections provenant du Yémen, qui a définitivement modifié l'échelle d'appréciation de l'emploi de papiers Abū Šubbāk. En effet, sur les 66 volumes[11] manuscrits contenant 111 livres datés[12], écrits aux xixe et xxe s., que l'on a pu examiner, 32 volumes, en fait 58 livres datés, ont été rédigés sur du papier Abū Šubbāk, soit 48,48 % des volumes et 52,25 % des livres.

Au Yémen, le papier Abū Šubbāk est parvenu certainement jusqu'à Zabid, dans la région de la Tihāma, qui forme une bande côtière le long de la mer Rouge. C'est ce que nous indique l'identité du copiste des m/ḥ 1, copié en 1375/1956, et 50/3, copié en 1358/1939–1940. Il a atteint plus tôt encore le Nord du pays, comme cela ressort des nombreux textes zaydites, en particulier ceux destinés à catéchiser (colophon des ms. UbL : Or. 25.663/3, copié « asfal madīnat Ṣaʿda » en 1338/1920, et Or. 25.744/3, copié en 1356/1937 « bi-balad Ṣaʿda »)[13]. Reste à mener une enquête complémentaire pour le Hadramaout.

Éthiopie

En 1997, le manuscrit BnF Ar. 7264, renfermant deux codex, était signalé comme copié en totalité sur du papier Abū Šubbāk[14]. La dimension maximale d'un feuillet est de 35,2 × 23,1 cm. Il est jaune d'apparence et ses bords ont bruni. Le premier texte, sans titre, formé de 3 feuillets, est daté au colophon de 1345/1926–1927. Le second livre, le *Kitāb Tanbīh al-anām wa-šifāʾ ilā saqqām fī bayān ʿulū Maqām nabīnā Muḥammad ʿalayhi afḍal al-ṣalā wa-azkā al-salām* de ʿAbd al-Ǧalīl b. Muḥammad b. Aḥmad Ibn ʿAẓẓūm al-Qayrawānī al-Murādī,

---

11 Un des volumes est en 2 tomes, sous la même cote, un autre en 3 tomes.

12 I. e. 31 vols. sont des collections. Le terme « livre » a été préféré à celui de « titre », qui ne figure pas toujours. Le décompte suit le système de catalogage de Voorhoeve, qui prend en compte les textes non-documentaires des *marginalia*, tels que les poèmes ou les *fawāʾid*. Mais quelques livres ont échappé au catalogage. « Daté », i. e. daté au colophon, ce qui ne signifie pas non-datable.

13 Grâce à l'historique des collections de l'UbL, on apprend que les manuscrits Or. 25.735 à Or. 25.746 ont été acquis en 1999 à Ṣaʿda et Manāḫa.

14 Mentionné par G. Humbert, « Le manuscrit arabe et ses papiers », 2002, 67. Voir la fiche du Catalogue en ligne de la BnF, http://archivesetmanuscrits.bnf.fr/ead.html?id=FRBNFEAD000038661&qid=sdx_q3.

a été achevé de copier, sur 185 feuillets, en ḏū al-qaʿada 1345/1927[15], soit la même année que le premier texte. Le manuscrit a circulé en milieu éthiopien, ainsi que l'indiquent une note en amharique[16], ainsi qu'une marque de propriété au nom de ʿUmar b. Abā Yasū de Limmū, suivie de la transmission du livre à ses enfants à la date de 1345 H, dont le texte reprend ponctuellement la formule de *waqfiyyas* de corans déjà notée pour l'Éthiopie[17], un milieu où le *K. Tanbīh al-anām* est extrêmement populaire[18]. À cette liste, s'ajoute la reliure plein cuir à rabat, avec son décor sur les plats et ses aies de 2 × 2 cartons. En ce qui concerne l'œuvre de copie du texte principal, on détecte des traits communs avec les manuscrits yéménites. Mais les dimensions du manuscrit, et, surtout, des éléments paléographiques, rendent plus probable le fait qu'il ait été copié en Éthiopie.

En Éthiopie, nous avons trouvé du papier Abū Šubbāk dans des manuscrits de la Bibliothèque de l'Institut d'études éthiopiennes (IES) de l'Université d'Addis Abeba et de la Bibliothèque du musée de Harar, à Harar City, appartenant à ʿAbdullah Sharif.

Une recherche systématique sur des codex en langue arabe des XIXe et XXe s. conservés à la Bibliothèque de l'IES, menée lors d'une mission préliminaire en 2010, suivie d'une seconde mission au printemps 2015, a montré que, sur 29 livres, 9 d'entre eux avaient été copiés sur ce papier, dont 6 datés au colophon (par ordre chronologique, IES 4624, 5505, 5517/1, 2637, 5509, 4574/2, voir tableau, section III.) et 3 non-datés au colophon (ms. IES 265, 4558, 4627)[19].

15 La date est écrite dans le sens de l'écriture arabe, de gauche à droite, contrairement à la normale, c'est-à-dire : 1, puis 3, puis 4 et enfin 5.

16 La note est écrite d'une écriture malhabile, à l'aide d'un stylo bille ou d'un crayon bleu clair qui a mal résisté au temps. Elle n'est donc pas aisée à déchiffrer. Y figure la date du 24 de ṭərr 1974 (calendrier éthiopien) qui correspond au lundi 1er février 1982 de notre ère. Je remercie Alessandro Gori de sa lecture.

17 A. Gori, « *Waqf* certificates of Qur'ans from Harar: A first assessment », 2015; A. Regourd, *Handlist of the Manuscripts in the ʿAbd Allah Sharif Museum*, à paraître, Annexe II. *Waqf* statements.

18 En juin 2016, 57 exemplaires du *K. Tanbīh al-anām* avaient déjà été identifiés dans la base de données de l'ERC « Islam in the Horn of Africa » (en cours, non encore publique), provenant de différentes régions du pays. Le manuscrit de la BnF y est décrit sous BnF 1, http://islhornafr.tors.sc.ku.dk/backend/manuscripts/1495.

19 Voir A. Gori, *A Handlist of the Manuscripts in the Institute for Ethiopian Studies. The Arabic Material of the Ethiopian Islamic Tradition*, 2014, IES 265, estimé du XXe s., 4 ; IES 2637, 35 ; IES 4558, estimé XXe s., 59 ; IES 4574/2, 64 ; IES 4624, 70 ; IES 4627, estimé XIXe s., 70–71 ; IES 5505, 74 ; IES 5509, 75 ; IES 5517/1, 77. Voir aussi la base de données de l'ERC « Islam in the Horn of Africa ». Au bas du dernier feuillet du ms. IES 4558, en dehors du cadre rectangulaire tracé autour de la fin du texte, on lit : « *fī sana* 1290 H », soit 1873–1874, mais

Des indications dont on dispose sur la provenance des fonds, la langue des textes, leurs auteurs et leur circulation, auxquelles s'ajoutent des caractéristiques paléographiques, permettent d'avancer, pour certains des codex, qu'ils ont été copiés en Éthiopie, pour les autres qu'ils auraient pu l'être. C'est le cas du ms. IES 265, dont le texte est en ancien harari[20]. Le Šayḫ Ḥabīb, scribe et propriétaire du ms. IES 5517, et propriétaire et auteur de notes en marges du ms. IES 4624, a vécu à Amūma (ou : Amūmu) dans le Qāllu-Wällo. L'auteur des textes des manuscrits IES 5517/1, *Muḫtaṣar Ḥilyat al-sāmiʿīn fī ḏikr šayʾ min aḫbār mabdaʾ sayyid al-šāfiʿīn*, et IES 4624, *Naṣīḥat al-aṭfāl wa-buġyat al-riǧāl*, Ǧawhar b. Ḥaydar b. ʿAlī al-Šonkī (m. 1937), a vécu à Shonkē dans le Sud-Est du Wällo. Le ms. IES 4627 a été la propriété du Ḥāǧǧ Sirāǧ Muḥammadī b. Muḥammadī al-Waqālawī. Dans l'historique des collections de l'IES, il est décrit comme faisant partie d'un lot acquis en 2006 auprès de quelqu'un du Wällo-Nord du Šäwa. Une lecture possible de la *nisba* du propriétaire pourrait être « Wuchalewi » de la ville de Wuchale (le Uccialli de la tradition coloniale italienne), mais l'usage d'un « *qāf* » pour écrire le « coup de glotte » du « *cha* » en amharique est habituellement transcrit par un *ṭ* avec trois points, en amharique *ʿaǧamī*[21]. Les textes des manuscrits IES 5505, le *Manhaǧ al-ṭullāb* d'Abū Yaḥyā b. Muḥammad b. Zakariyyāʾ Zayn al-Dīn al-Sunaykī al-Šāfiʿī al-Anṣārī (ixe/xve s.) et IES 5509, *ʿUmdat al-sālik wa-ʿiddat al-nāsik* d'Aḥmad b. Luʾluʾ b. al-Naqīb Šihāb al-Dīn al-Rūmī al-Miṣrī al-Šāfiʿī (m. 1368/1948–1949), sont des textes chafiites, dont on dénombrait 13 exemples pour chacun d'eux dans la base de données de l'ERC « Islam in the Horn of Africa » en juin 2016. Autres textes fréquents, ceux contenus dans les manuscrits IES 2637, le *Muḫtaṣar* du *Tanbīh al-ānām*, déjà rencontré, IES 4627, la *Mulḥat al-iʿrāb* du très fameux auteur de *Maqāmāt*, Abū Muḥammad al-Qāsim b. ʿAlī b. Muḥammad b. ʿUṯmān b. al-Ḥarīrī al-Baṣrī (2e moitié du ve/xie s.), dont on comptait 7 exemples au total dans la même base de données à la même date, enfin, l'IES 4558, *Muḫtaṣar Abī Šuǧāʿ ʿalā maḏhab al-Šāfiʿī* (connu sous d'autres titres) d'Abū Šuǧāʿ Aḥmad

la main semble différente et l'encre plus fraîche ; une lecture par 1390 donnerait 1970–1971, une date qui nous paraît trop récente pour être celle de la copie ; on s'est rangé à l'estimation d'A. Gori qui place la copie au xxe s., une estimation en concordance avec le nuage statistique des papiers traités ici, voir note suivante.

20 Il s'agit du *Kitāb al-farāʾiḍ li-al-ʿArāšī*, i. e. ʿAbd al-Raḥmān al-ʿArāšī, sur la biographie de qui nous ne disposons d'aucune information. Voir note 17, et pour le fonds harari de la Bibliothèque de l'IES, A. Gori, *A Handlist of the Manuscripts in the Institute for Ethiopian Studies*, 2014, xli.

21 Alessandro Gori, communication personnelle, 06.07.2016.

b. Ḥasan/Ḥusayn b. Aḥmad Taqī al-Dīn Aḥmad b. al-Ḥusayn al-Iṣfahānī (né en 434/1042–1043, en vie en 500/1106–1107), en 4 exemplaires toujours dans la même base de données, en juin 2016. Les marges du *Šarḥ muṯallaṯ Quṭrub* de Šams al-Dīn Abū al-Qāsim ʿAbd al-Wahhāb b. al-Ḥasan b. Barakāt al-Muhallabī al-Warrāq al-Bahīʾī (m. 1278/1861–1862 ; IES 4574/2) n'apportent rien de particulier en termes de circulation des textes.

En 2010, 5 occurrences étaient relevées sur 16 manuscrits pris au hasard dans le fonds du Sherif Museum, Harar City. Ce fonds s'élève à plus de 500 manuscrits, toutes dates confondues[22].

Les manuscrits 74 et 59 de la collection privée Gerald et Barbara Weiner (Chicago), consultée au printemps 2014, ont été également écrits sur ce papier. Le ms. 74 contient la *Mulḥat al-iʿrāb*, déjà rencontrée. Seul le ms. 59, *Al-muqaddima al-ḥaḍramiyya fī fiqh al-Sāda al-šāfiʿiyya*, où à nouveau l'école chafiite est concernée, est daté au colophon, qui, en fait porte deux dates, 1313/1895–1896 et 1363/1943–1944. Le sujet renvoie à la grande région sunnite du Sud du Yémen, le Hadramaout. Ces deux manuscrits pourraient avoir été copiés en Éthiopie.

Autres pays

En dehors du Yémen et de l'Éthiopie, une étude s'appuyant sur des sources scripturaires, mentionne, dès la fin des années 60, l'existence d'un papier *abu shibbāk* parmi les importations européennes en direction du Soudan de l'Est, pour la fin du xixe s.[23]. Aucune trace n'a été relevée en Égypte à ce jour, mais les études sur les papiers filigranés publiées, connues de nous, s'arrêtent à la fin du xixe s.[24]. Il est intéressant de noter qu'aucun papier Abū Šubbāk n'est apparu

---

22 Le chiffre de 500 correspond aux manuscrits numérisés par Abdullah Sharif en 2009–2010. Ils n'ont pas reçu de cote du musée. La base de données de l'ERC « Islam in the Horn of Africa » inclut les manuscrits numérisés du musée de Harar. Le Harar est un haut lieu de la copie de manuscrits arabes en Éthiopie, cf. R. S. O'Fahey, *The Writings of the Muslim Peoples of Northeastern Africa* (*Arabic Literature of Africa*, vol. 3, fasc. A), section 1, The Near and Middle East, 2003, vol. 13, 20, sur le fonds de manuscrits arabes provenant du Harar.

23 N. Shuqayr, *Ǧuġrafiyya wa-tārīḫ al-Sūdān*, Beyrouth, 1967, 177, d'après T. Walz, « The Paper Trade of Egypt and the Sudan in the Eighteenth and Nineteenth Centuries », 1985, 29–48, 39–40. Repris par J. M. Bloom, « Paper in Sudanic Africa », 2008, « At the end of the nineteenth century, a thick brown (perhaps cream colored) paper called *abu shibbāk*, presumably referring to its weave or even texture, was also imported from Europe », 54.

24 En dehors de l'étude de Walz qui s'interrompt en 1880 (voir *infra*), il s'agit de : ʿI. A. ʿĪsāwī, *Al-waraq fī Miṣr fī al-qarn al-tāsiʿ ʿašar wa-ahamiyyatuhu fī al-dirāsāt al-diblūmāṭiyya wa-al-bibliūġrāfiyya*, [2002], 58 filigranes relevés à main levée, dans deux annexes de son

à ce jour au Nigéria alors qu'en 2014, l'étude de Michaelle Biddle (Wesleyan University), centrée sur le papier, portait déjà sur ca. 1.250 manuscrits – incluant des pièces du xxe s. – et 65.000 f.

Le papier Abū Šubbāk a été signalé en Extrême-Orient en janvier 2010, tout d'abord, dans un manuscrit de *fiqh* provenant de la Regency d'Aceh Besar, au Nord de Sumatra, acéphale et amputé de la fin du texte ; puis, entre janvier et octobre 2010, dans un (plusieurs ?) codex conservé(s) en Indonésie. Il s'agit de manuscrits en langue arabe qui, selon l'état de la recherche, ne seraient pas antérieurs au xixe s.[25]. Ces manuscrits ayant pu être importés, en particulier du Yémen et plus précisément du Hadramaout, il serait imprudent pour l'instant d'en conclure à la circulation du papier Abū Šubbāk dans la zone concernée[26].

Ce papier a été utilisé pour des livres sans distinction de sujet : corans (UbL Or. 25.179), textes religieux (*ḫuṭba*, UbL Or. 6330, récitation du Coran, FHS 97, UbL Or. 6362/2, « catéchisme » zaydite, UbL Or. 6637, prières, UbL Or. 25.174/1, traité zaydite pour la *daʿwa*, UbL Or. 6362/4, dogmatique zaydite, UbL Or. 23.268), théologie dialectique (*kalām*, UbL Or. 25.287/2, Or. 6643/3), droit casuistique et ses fondements (*fiqh*, m/ḥ 66/1, UbL Or. 6368, *uṣūl al-fiqh*, UbL Or. 6634, Or. 6643/2), *farāʾiḍ* (UbL Or. 6643/1), soufisme (UbL Or. 25.287/3, UbL Or. 23.407/11), histoire (Zabid m/ḥ 1 ; UbL Or. 6357a), grammaire (UbL Or. 25.288/1 et 3, *ṣarf*, UbL Or. 23.407, UbL Or. 25.174/6), *adab* (UbL Or. 6362/1) et poésie (UbL Or. 23.447/1, UbL Or. 25.174/4), divination (UbL Or. 23.407/4).

étude des papiers de documents et registres (*siǧillāt*) du xixe s., 94–122 ; A. M. Ṣ. Ibrāhīm, *Al-ḫuṭūṭ wa-al-ʿalāmāt al-māʾiyya fī al-maḫṭūṭāt al-ʿarabiyya*, 2010, papiers filigranés de documents arabes de la Bibliothèque des Waqfs du Caire, dont le plus récent date de 1303/1885–1886.

25 Russel Jones, "Indonesia and Malay Manuscripts and links to the Ottomans", 2010, en ligne.

26 Sur l'émigration de lettrés Hadhramis en Indonésie, en général, les travaux sont nombreux. On se contentera de citer ici : R. B. Serjeant, *The Sayyids of Hadhramaut*, 1967 ; *id.*, « History and Historiography of Hadhramaut », 1962, 239–471, nombreuses références, entre autre 255 ; différents articles dans U. Freitag et W. G. Clarence-Smith, *Hadhrami Traders, Scholars and Statesmen in the Indian Ocean, 1750s–1960s*, 1997. En ce qui concerne le nord de Sumatra, U. Freitag mentionne le cas, considéré comme exemplaire, de quatre Sayyids hadhramis de Tarīm, qui ont émigré dans la province d'Aceh (Aceh Province), où se trouve le Regency d'Aceh Besar, à la fin du xviiie s., « Reflections on the Longevity of the Hadhrami diaspora in the Indian Ocean », 2009, 17–32, particulièrement, 23*sq.* et note 18. Abdul R. Tang Abdullah, « Arab Hadhramis in Malaysia: their origins and assimilation in Malay society », 2009, 47*sq.*, insiste, à partir du cas de la Malaisie de la fin du xviie s., sur la double activité des Sayyid-s du Hadhramaout, pratiquant le commerce en s'appuyant sur leurs réseaux commerciaux, d'une part, clamant leur rôle religieux, d'autre part : on a donc là un réseau de circulation potentiel d'ouvrages qui perdure aux xviiie et xixe s.

Il a servi à différents formats :

Format rectangulaire (ex. tirés de l'UbL),
- petits [H de 11,5 à 12,8 cm × l 8,5 à 9 cm],
- type « cahiers d'écoliers » [H de 23,3 à 25,2 cm × l de 16,2 à 18,5 cm],
- grands, [tels le UbL Or. 23.449 : H 35,5 × l 24,5 cm],
- et « intermédiaires » [H de 17,5 à 18,3 cm × l de 11,7 cm à 13),

enfin, format *safīna* (BnF Ar. 7084, non daté, antérieur à 1922, 24,5 × 18 cm)[27].

Méthodologie

La méthodologie suivie est identique à celle décrite dans l'introduction à ce volume. Inspirée par l'article de Terence Walz publié en 1985[28], elle a été mise au point au cours de notre étude des papiers Abū Šubbāk. Elle s'appuie sur les papiers comme source primaire de l'histoire du commerce et, en première instance, de la reconstitution de routes commerciales.

Première proposition de classification des papiers Abū Šubbāk : papiers *Beyāḍ Abū Šubbak* et *Waraq Abū Ši/ubbāk*

Nous reprenons ici la classification établie dans notre étude précédente[29], avec des déterminations reprécisées pour chaque type de papier du fait d'un corpus étoffé de manuscrits conservés au Yémen et en Éthiopie.

Les papiers de type A donnés en illustration sont empruntés d'abord aux manuscrits les plus anciens datés, 1327/1908–1909 et 1324/1906 (papier A.2.) ; ces papiers sont faits à la machine, au cylindre. De même, c'est le papier le plus ancien de type C qui est illustré. Un seul cas de papier B est connu. Aucun des quelques exemples connus de type D n'est daté. Les papiers B, C et D rencontrés sont des papiers faits à la machine, B et D, au cylindre.

Notre classification en différents types repose principalement sur l'observation à l'œil nu des marques du papier. Lors de l'enquête, d'autres différences, nombreuses, se sont fait jour entre les papiers de type A, d'ordre sensible

27 Voir J. Dufour & A. Regourd, « Les *safīnas* yéménites », à paraître.

28 T. Walz, « The Paper Trade of Egypt and the Sudan in the Eighteenth and Nineteenth Centuries », 1985.

29 A. Regourd, « Paper Trade on the Red Sea (XIXe c.–first half of the XXe c.): 'local papers' against Italian's. The case of an 'Ottoman' watermarked paper used in Yemen and in Ethiopia », à paraître.

(apparence, couleur, touché, ...). Elles résultent certainement de la composition de la pâte. Une analyse en laboratoire a d'ores et déjà été menée sur l'un des papiers, soyeux au toucher (voir papier B). Mais l'enquête menée à l'UbL montre une très grande variété dans l'apparence des papiers et dans leur évolution au contact d'environnements multiples sur la durée. Ces observations à l'œil nu, que nous avons dûment consignées, font voir la nécessité d'une seconde recherche par analyse en laboratoire. Elle permettrait de décliner des fourchettes de datation plus précises, appréciables dans la mesure où le type A est le plus courant, et, aussi, d'affiner les hypothèses de localisation pour la fabrication de ces papiers[30]. L'intérêt de la classification actuelle (types A, C et D) réside dans le fait que, ne requérant pas de spécialisation particulière, elle puisse être utilisée largement.

Les marques prises en compte sont : les vergeurs, les fils de chaînette, les filigranes et les contremarques. Au sein du type A, des différences graphiques apparaissent entre les filigranes, ainsi que des variantes entre les contremarques dans la composition (organisation du texte) ou l'écriture. Là encore il serait possible de raffiner la typologie : les sous-groupes qui en ressortiraient, qu'ils soient synchroniques ou diachroniques, permettraient d'aboutir à des fourchettes de datation plus précises.

Ont été identifiés à ce jour deux grands groupes de papiers Abū Šubbāk, les papiers *Beyāḍ Abū Šubbak* et les papiers *Waraq Abū Ši/ubbāk*.

Les papiers *Beyāḍ Abū Šubbak*

Le premier groupe de papier est de loin celui qui repose sur la base statistique la plus importante à ce jour.

Type A

Papier filigrané, arborant un croissant à profil humain inséré dans un écu à double contour et portant la contremarque :

[1re ligne] *Beyāḍ Abū Šubbak isṭambūlī* (بياض ابوُ شبّك اسطمبولي)
[2e ligne] *ʿālī aṣīlī* (عالي اصيلي)[31]

C'est de loin le plus fréquent au Yémen de même qu'en Éthiopie.

30 Une recherche de fonds est en cours.

31 Mêmes observations que dans la note 25.

Type A.1

A.1. diffère d'A.2. par les dimensions de l'écu qui contient le croissant à profil humain, ainsi que par la longueur occupée par la 2e ligne de la contremarque. Une autre différence très utile à un repérage rapide du type du papier réside dans une sorte de « défaut » situé dans le dernier terme de la contremarque, « *aṣīlī* (اصيلي) », présent en A.1. et absent de A.2. : la ligature entre le premier *yāʾ* et le *lām* n'est que partiellement présente. Étant donné la fréquence de ce défaut, il est peu probable qu'il résulte de l'adhérence de la pâte, mais provient plutôt de l'installation de la contremarque sur la trame [Ill. 4.1b, et Annexe].

Les dimensions des différentes marques du papier, prélevées sur les manuscrits de l'UbL, de Zabid et de l'IES, sont les suivantes[32] :

a. entre 2 fils de chaînette : 2,5 cm ;
b. pour 20 vergeurs : 3 à 3,2 cm ;
c. H maximale de l'écu (entre les pointes supérieure médiane et inférieure) : 6,3 à 6,7 cm ;
d. distance entre les deux pointes supérieures à dextre et sénestre de l'écu : 4,7 à 4,9 cm ;
e. l maximale de l'écu (partie renflée) : 4,8 à 5 cm ;
f. L maximale de la ligne supérieure de la contremarque : 13,5 à 14 cm ;
g. L maximale de la ligne inférieure de la contremarque : 5,8 à 6,1 cm.

Il faut compter avec une marge d'erreur dans les relevés des mesures. Cependant, prélevées sur une quantité importante, un tassement se produit autour de fourchettes données. Cela permet de poser l'existence de différents rouleaux, ici et pour les types suivants.

32 Les dimensions n'ont pas été prises systématiquement sur tous les feuillets. Les mesures entre 2 fils de chaînettes sont prises du côté interne de la marque. La distance entre les deux pointes supérieures à dextre et sénestre de l'écu sont prise à l'aplomb de la pointe. Toutes les autres mesures sont prises à partir du côté externe de la marque. La longueur des textes de la contremarque cherche à être maximale, elle est donc prise à partir de la partie la plus avancée de la première et de la dernière lettre. Il n'est pas toujours possible de réunir les 7 mesures sélectionnées pour chaque papier : filigranes et contremarques apparaissent fréquemment dans la pliure du bifeuillet, partiellement sur petits formats, voire pas du tout, et, parfois, les marques sont évanescentes. Le ms. UbL Or. 23.448 a des mesures sensiblement différentes pour la largeur de l'écu et, sur quelques feuillets, pour la longueur du texte de la contremarque ; les Or. 25.745 et Or. 23.407/10 ont une hauteur d'écu plus importante (le second sur très peu d'occurrences) ; sur un nombre limité d'exemples, la longueur du texte inférieur de la contremarque en Or. 23.407/6–7 est plus importante.

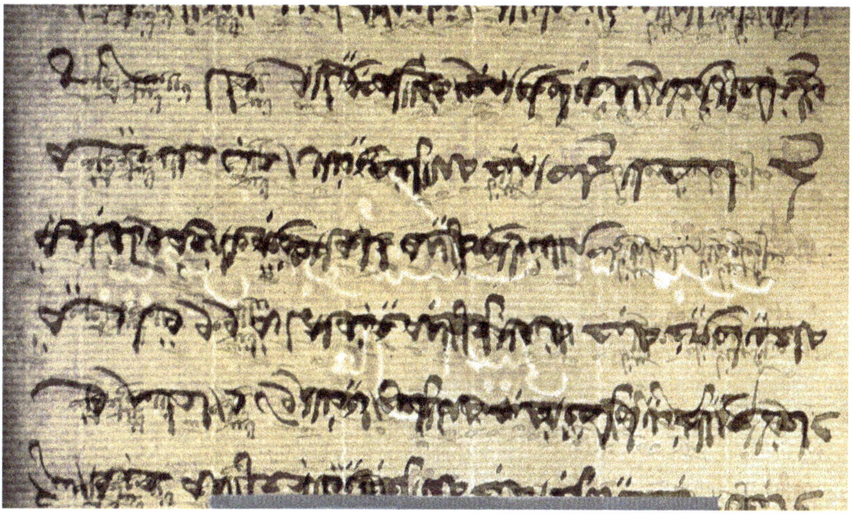

ILLUSTRATION 4.1A *Type A.2. Contremarque. Ms. W41/EMIP 00543, daté 1327/1908–1909*
© AIMABLE AUTORISATION EMIP

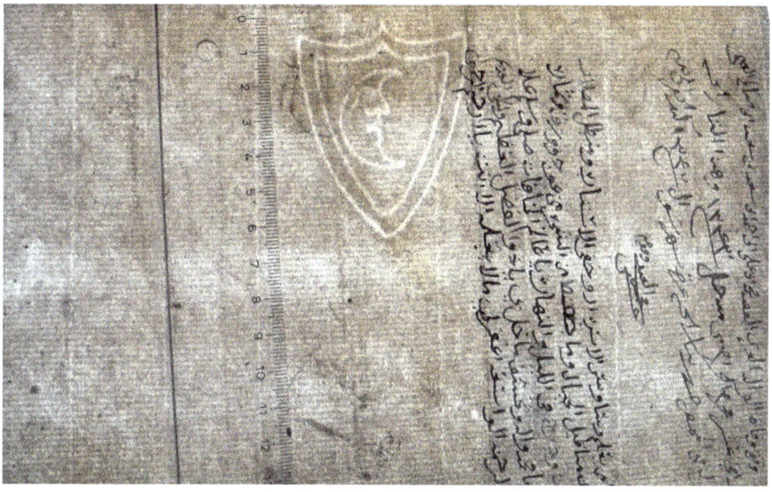

ILLUSTRATION 4.1B *Type A.1. Écu. Ms. UbL Or. 6330, daté 1346/1927*
© AIMABLE AUTORISATION UBL

Le type A.1. est présent au Yémen de 1345/1926 à 1375/1956, c'est-à-dire sur une période de trente ans, et en Éthiopie, sur une base statistique encore réduite, de 1328/1910 à 1355/1936, soit sur une période d'une trentaine d'année, avec peut-être une pointe jusqu'en 1363/1943–1944.

Type A.2

Il diffère du type A.1. par les dimensions de l'écu contenant un croissant à profil humain, de même que par la longueur de la contremarque, surtout sensible à la seconde ligne. Il ne comporte pas le défaut signalé pour A.1., situé dans le dernier terme de la contremarque (« *aṣīlī* اصيلي ») [Ill. 4.2a–e et Annexe].

Les dimensions des différentes marques du papier sont les suivantes (même corpus que pour A.1.) :

a. entre 2 fils de chaînette : 2,5 cm ;
b. pour 20 vergeurs : 2,8 à 3,2 cm ;
c. H maximale de l'écu (entre les pointes supérieure médiane et inférieure) : 6,8 à 7,4 cm ;
d. distance entre les deux pointes supérieures à dextre et sénestre de l'écu : 4 à 4,6 cm ;

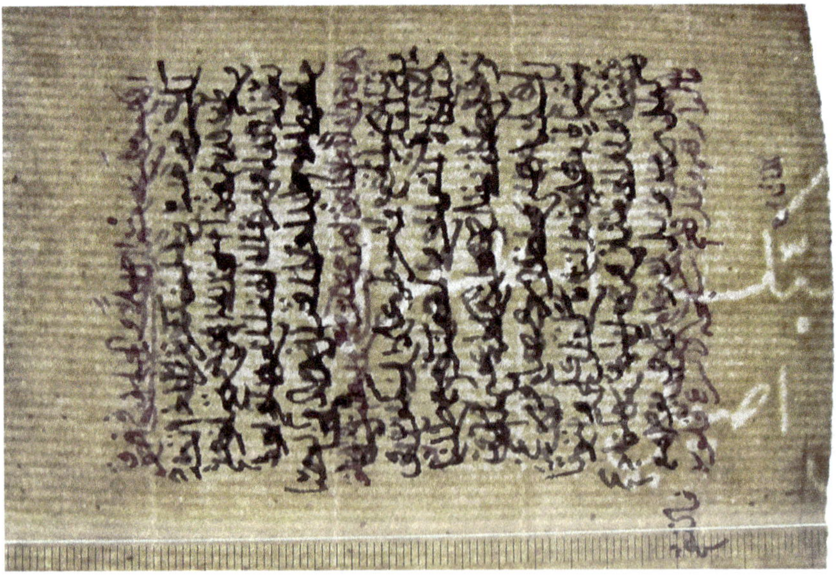

ILLUSTRATION 4.2A *Type A.2. Contremarque. Ms. UbL Or. 25.288/3, daté 1324/1906* © AIMABLE AUTORISATION UBL

ILLUSTRATION 4.2B *Type A.2. Écu. Ms. UbL Or. 6362/2, daté 1328/1910*

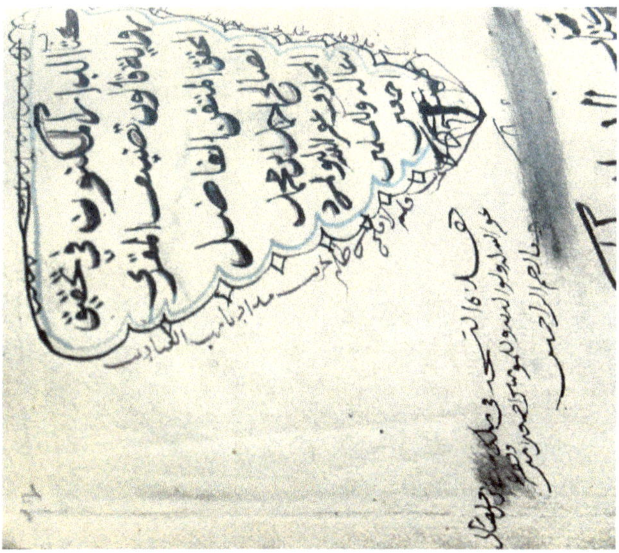

ILLUSTRATION 4.2C *Type A.2. Apparence externe. Page de titre. Ms. UbL Or. 6362/2, daté 1328/1910*

ILLUSTRATION 4.2D *Type A.2. Contremarque. Ms. UbL Or. 6362/2, daté 1328/1910*

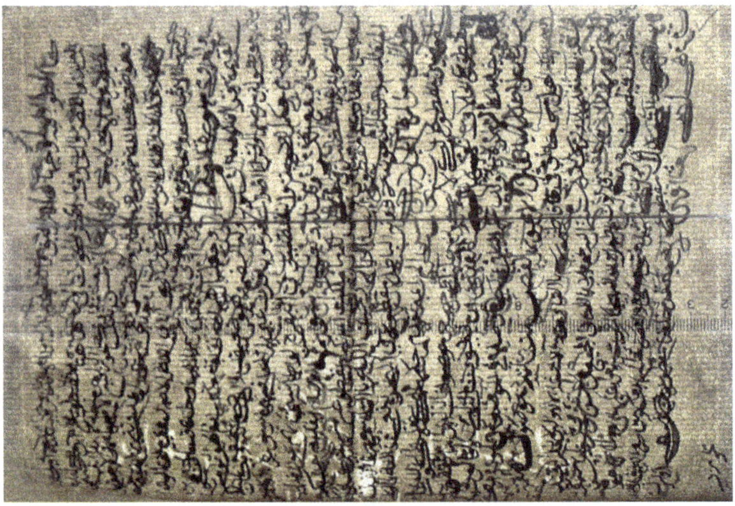

ILLUSTRATION 4.2E *Type A.2. Contremarque. Ms. UbL Or. 6362/2, daté 1328/1910*

e. l maximale de l'écu (partie renflée) : 4,3 à 4,9 cm ;
f. L maximale de la ligne supérieure de la contremarque : 13,3 à 13,8 cm ;
g. L maximale de la ligne inférieure de la contremarque : 4,6 à 5 cm[33].

Le papier A.2. a été utilisé au Yémen sur près de 120 ans, de 1263/1846–1847 à 1386/1967. Le nuage statistique dégagé pour le Yémen situe nettement l'usage de ce papier entre 1324/1906 et 1386/1967, soit sur une période d'une soixantaine d'années. Une seule occurrence signale son emploi au xixe s., et, surtout, elle est suivi d'un vide remarquable : alors que le papier A.2. est attesté continûment à partir de 1324/1906 jusqu'à la date extrême de la fourchette de son emploi, rien n'a l'air de se passer entre 1263/1846–1847 et 1324/1906. Cela jette en retour un doute sur le colophon du ms. 60, dont la date a peut-être été mal lue ou notée par les catalogueurs, ou bien est fausse dans le texte manuscrit, une donnée que nous n'avons actuellement pas le moyen de vérifier.

En Éthiopie, on le relève de 1327/1908–1909 à 1353/1934–1935, soit 25 années environ, mais sur une base statistique très faible.

A.1., A.2.1., A.2.2. etc.

Sur certains papiers A.2., le point diacritique du *bāʾ* de *Bayāḍ*, à la ligne supérieure de la contremarque, apparaît très excentré (mss Zabid m/ḥ 59/9, 60, UbL Or. 6353a, b ; Or. 6362/1, Or. 6362/2 ; Or. 6368 ; Or. 6634 ; Or. 23.268 ; Or. 25.287/3 ; Or. 25.662, f. 189–238, 243–244, 247–252 ; Or. 25.663/3). Autre particularité de certains papiers, les différents mots formant la contremarque apparaissent sous forme de blocs distincts : l. supérieure : 1. *Bayāḍ* ; 2. *Abū Šubbak* ; 3. *isṭambūlī*, séparés par 1 cm ; l. inférieure : *ʿālī* et *aṣīlī*, séparés par 0,9 cm (par ex. ms. Or. 6353b). Enfin, on observe des différences dans la forme du *ʿayn* ou du *yāʾ* de *ʿālī*, ainsi que dans leur installation, variable, sur les fils de chaînettes. Ces différences se superposent à celle observée entre les contremarques de A.1. et A.2, venant de l'absence ou non de ligature, au dernier mot de la ligne inférieure. Elles – surtout les variations touchant à la forme des lettres et à leur installation sur la grille – suggèrent plusieurs cylindres à l'œuvre au même moment.

On détecte des différences entre les papiers regroupés sous le type A.2. sur la durée. Les facteurs de l'environnement et de la conservation sont bien sûr

---

33 Différences : H écu : IES 5509, Or. 6643 ; L de l'une ou des deux lignes de la contremarque : IES 5509, Or. 6353b, Or. 6634, Or. 23.268 (peu d'occurrences), Or. 23.448 (un certain nombre d'occurrences).

à prendre en compte dans l'évolution des papiers. A.2., bien qu'utilisé pour la production de codex, paraît d'une moins grande qualité.

Le ms. UbL Or. 6634, daté au colophon du 24 ḏū al-ḥiǧǧa 1346/13.06.1928 a été entièrement copié sur des papiers A.1. et A.2., formés en quaternions, avec une majorité de papiers A.1. Les cahiers ayant été laissés libres, l'observation en un clin d'œil de l'ensemble empilé montre que ces cahiers n'ont pas la même épaisseur, signe de l'épaisseur elle-même inégale des papiers [ill. et commentaire Annexe].

Le papier du Catalogue d'al-Ḫizāna al-ʿāmira[34] *: Papier B*

Al-Ḫizāna al-ʿāmira est une bibliothèque établie par l'Imam Yaḥyā b. Muḥammad b. Yaḥyā Ḥamīd al-dīn (1869–1948) dans la Grande mosquée de Sanaa. Son Catalogue, le *Fihrist kutub al-ḫizāna al-mutawakkiliyya al-ʿāmira bi-al-ǧāmiʿ al-muqaddas bi-Ṣanʿāʾ al-muḥammiyya*, a peut-être été imprimé plusieurs fois [ill. 4.3a][35]. L'exemplaire en notre possession ne porte pas de date, mais l'une des décisions prises par la cour de l'Imam, figurant en fin de catalogue, permet d'en estimer l'impression à 1361/1942–1943 au plus tôt[36]. Il provient des presses typographiques yéménites de l'Imam Yaḥyā ; l'Imam disposait d'une petite presse à l'Office de l'éducation (Idārat al-maʿārif) pour les livres et, à la cour (*maqām*), d'une ancienne presse laissée derrière eux par les Ottomans sur laquelle était imprimé l'unique journal autorisé, un mensuel, puis un quotidien, *al-Īmān*[37].

D'une manière générale, il est intéressant de noter, au travers de l'exemplaire en notre possession, que le livre imprimé a été traité comme s'il s'agissait d'un manuscrit. Tout d'abord, un cadre noir délimite la surface d'écriture. Si l'on en croit ce qui subsiste de la reliure, une pièce de tissu à impressions colorées a été apposée sur le bloc de cahiers et est maintenue par une bande de papier encollée, par-dessus laquelle une tranchefile formée de fils de deux couleurs

---

34 Je voudrais remercier Jean-Louis Estève (Paris, École Estienne) pour ses commentaires sur ce papier.

35 *Fihrist kutub al-ḫizāna al-mutawakkiliyya al-ʿāmira bi-al-ǧāmiʿ al-muqaddas bi-Ṣanʿāʾ al-muḥammiyya*, s. d.
«فهرست كتب الخزانة المتوكلية العامرة بالجامع المقدس بصنعاء المحمية»، صنعاء، وزارة المعارف المتوكلية، د. ت.

36 P. 328 du Catalogue.

37 Sur les presses yéménites au temps de l'Imam Yaḥyā Muḥammad Ḥāmid al-Dīn (r. 1918–1948), cf. E. Rossi, « La Stampa nel Yemen », 1938 ; R. B. Serjeant & R. Lewcock, *Ṣanʿāʾ. An Arabian Islamic City*, 1983 [1re imp.], chap. 8, « The Post Medieval and Modern History of Ṣanʿāʾ and the Yemen, ca. 953–1382/1515–1962 », 101a, 102a ; B. Messick, *The Calligraphic State. Textual Domination and History in a Muslim Society*, 1992, p. 118–119.

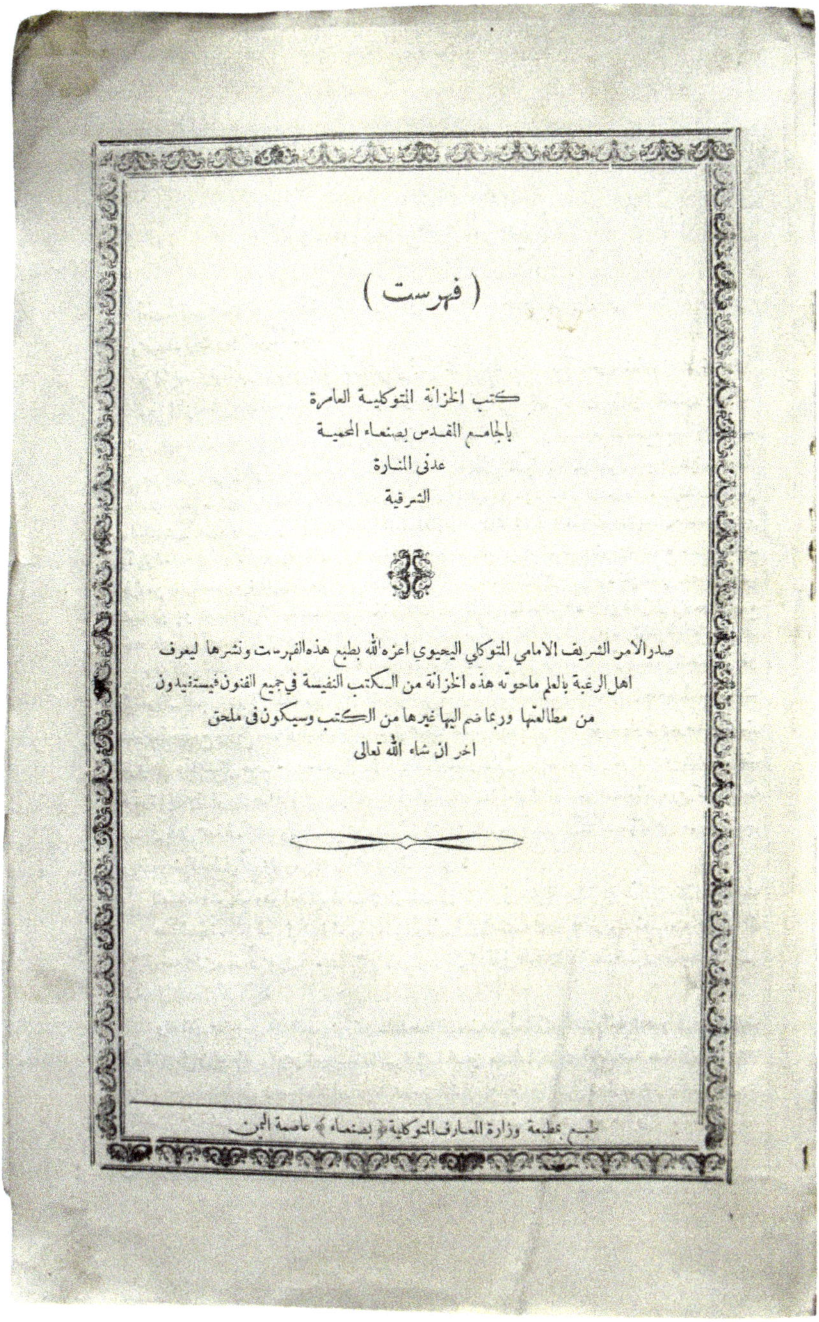

(فهرست)

كتب الخزانة المتوكلية العامرة
بالجامع المقدس بصنعاء المحمية
عدنى المنارة
الشرقية

صدر الامر الشريف الامامي المتوكلي اليحيوي اعزه الله بطبع هذه الفهرست ونشرها ليعرف
اهل الرغبة بالعلم ماحوته هذه الخزانة من الكتب النفيسة في جميع الفنون فيستفيدون
من مطالعتها وربما ضم اليها غيرها من الكتب وسيكون فى ملحق
اخر ان شاء الله تعالى

طبع بمطبعة وزارة المعارف المتوكلية ﴿ بصنعاء ﴾ عاصمة اليمن

ILLUSTRATION 4.3A Fihrist kutub al-ḫizāna al-mutawakkiliyya al-ʿāmira bi-al-ǧāmiʿ al-muqaddas bi-Ṣanʿāʾ al-muḥammiyya, *au plus tôt 1361/1942–1943* (rouge et blanc) a été cousue. C'est le papier de cette impression de (ou : postérieur à) 1361/1942–1943 que l'on se propose d'étudier[38].

Les marques du papier du catalogue *Fihrist kutub al-ḫizāna al-mutawakkiliyya* correspondent au type A.1., mais il s'agit d'un papier qui a été utilisé pour des impressions typographiées. Il a en conséquence fait l'objet d'une analyse de la pâte, qui permet une caractérisation d'ensemble du papier et une entrée séparée, (B), lui a été aménagée, où il n'a pas été qualifié de « type ».

Il s'agit d'un papier à motif de croissant à profil humain inséré dans un écu à double contour et portant la contremarque :

> « [1re ligne] *Beyāḍ Abū Šubbak isṭambūlī* (بياض ابوُ شبّك اسطمبولي)
> [2e ligne] *ʿālī aṣīlī* » (عالي اصيلي)

Un bifeuillet détaché permet d'observer que le croissant est tourné vers la droite. L'espace entre deux fils de chaînette, mesuré entre les deux bords internes, est de 2,4 cm, tandis que l'espacement entre 20 vergeures est de 3 cm, entre deux bords externes. Le papier a une apparence légèrement jaune [ill. 4.3b–c].

Des traces sous forme d'une série de traits courts et obliques sur tout le long du bord supérieur des feuillets, apparaissent régulièrement dans ce papier [ill. 4.3d, e]. Elles suggèrent qu'il était fabriqué à l'aide d'un cylindre, sur lequel était installée la trame : elles constituent le système de raboutage des deux bords de la trame métallique. Les fils de chaînette sont installés perpendiculairement à l'axe du cylindre. Les premiers cylindres pour la production de papier filigrané (ou Dandy roll) connus datent de 1826 et ont ensuite évolué[39].

La distance entre deux traces consécutives laissées par le raboutage des deux bords de la trame métallique donne la circonférence du cylindre. La dimension d'un bifeuillet du Catalogue est de 36,4 × 46,9 cm (H × l). Le filigrane est situé dans la partie droite du bifeuillet et la contremarque dans la partie gauche, la première ligne du texte étant placée légèrement plus haut que la partie supérieure du filigrane (ca. 2 cm). Sur chaque feuillet de l'ouvrage, apparaît soit le filigrane, soit la contremarque, i. e. aucun feuillet n'est vierge de l'une ou l'autre marque. La trace de raboutage, lorsqu'elle apparaît, est située soit au nord du filigrane et de la contremarque, soit au sud ; dans les autres

38 L'ensemble de cette étude a été réalisé sur mon exemplaire personnel.

39 D. Hunter, *Papermaking. The History and Technique of an Ancient Craft*, 1978 [repub. of the second, revised and enlarged 1947 ed.], 400. Voir également F. Didot, *Le Centenaire de la machine à papier*, 1900, qui attribue à un ouvrier français, Louis Nicolas Robert (1761–1828), employé à la papèterie d'Essonne, près de Paris, l'invention de la fabrication du « papier continu » en « papier à la mécanique », p. 30.

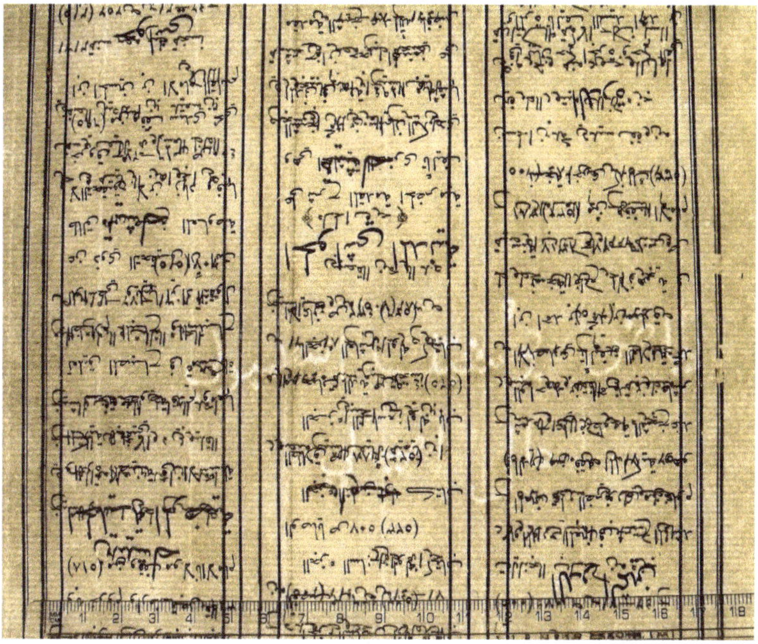

ILLUSTRATION 4.3B *B, contremarque. Catalogue* Fihrist kutub al-ḫizāna al-mutawakkiliyya, *au plus tôt 1361/1942–1943*
© PHOTOGRAPHIE ANNE REGOURD

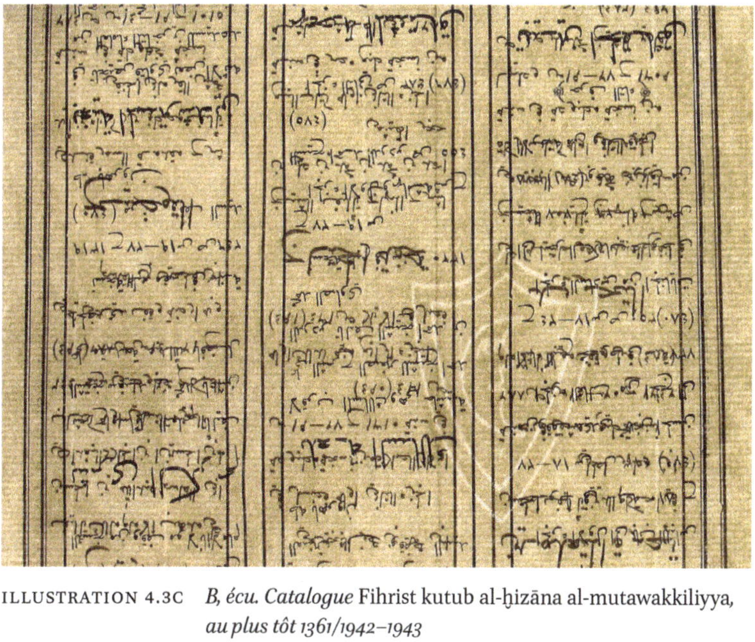

ILLUSTRATION 4.3C *B, écu. Catalogue* Fihrist kutub al-ḫizāna al-mutawakkiliyya, *au plus tôt 1361/1942–1943*
© PHOTOGRAPHIE ANNE REGOURD

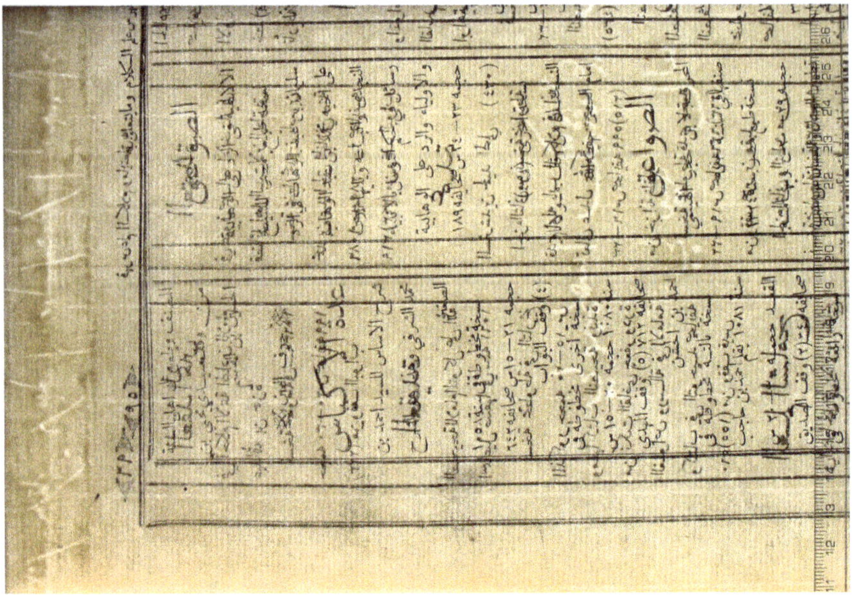

ILLUSTRATION 4.3D *B, contremarque et raboutage. Catalogue* Fihrist kutub al-ḫizāna al-mutawakkiliyya, *au plus tôt 1361/1942–1943*
© PHOTOGRAPHIE ANNE REGOURD

ILLUSTRATION 4.3E *B, écu et raboutage. Catalogue* Fihrist kutub al-ḫizāna al-mutawakkiliyya, *au plus tôt 1361/1942–1943*
© PHOTOGRAPHIE ANNE REGOURD

cas, le filigrane et la contremarque sont présents sans cette trace. Cela permet d'avancer que la circonférence du cylindre est approximativement de 36,4 × 3 = 109,2 cm, ce qui donne un diamètre d'environ 35 cm.

En outre, on note la présence de séries de pointillés sur un nombre conséquent de feuillets, parfois accompagnés d'un bâtonnet oblique. Ces marques ont été imprimées dans la pâte encore humide, de la même façon que le filigrane, la contremarque et les marques de raboutage. Elles sont très proches du bord du feuillet – inférieur / supérieur – du livre et parallèles à lui ou bien se confondent avec lui. Elles partent toujours du bord externe du feuillet et n'apparaissent que sur une courte distance, tout au plus 2 cm [ill. 4.3f]. Elles pourraient constituer un pré-découpage du rouleau de papier en feuillets. Sur l'un des bords, supérieur ou inférieur, d'un certain nombre de feuillets, on aperçoit la marque du raboutage et sur le bord opposé, les pointillés (voir par ex. p. 46). Si notre lecture de la fonction des pointillés était juste, elle confirmerait que, sur la longueur de papier filigrané produite par un tour de rouleau, on peut prélever plus d'un feuillet. En outre, ces marques seraient un indice de plus

ILLUSTRATION 4.3F *B, contremarque et marques de découpage. Catalogue* Fihrist kutub al-ḫizāna al-mutawakkiliyya, *au plus tôt 1361/1942–1943*

d'une production mécanisée de ce papier produit au cylindre. Enfin, le format du Catalogue serait celui d'un feuillet pré-découpé.

La pâte a fait l'objet d'une analyse par microscopie optique[40]. Il s'agit d'une pâte chimique blanchie à base de résineux, avec une très forte proportion d'épicéa, et de feuillus, en particulier du peuplier et du hêtre. On note la présence de plantes annuelles (paille, lin, chanvre) à l'état de fortes traces. La pâte a subi un traitement mécanique, perceptible en terme de coupe des fibres, fibrillation, gonflement de la paroi fibreuse et par la présence de nombreux éléments fins. L'endommagement de la paroi fibreuse a rendu l'identification des essences difficile à réaliser. Le papier est soyeux au toucher.

Les papiers *Waraq Abū Šibbāk/Šubbāk*

Le deuxième groupe est, au contraire, le moins illustré des deux à ce jour.

Type C. Le papier Waraq Abū Šibbāk al-aṣīlī

Les manuscrits m/ḥ 35 et 50 de la Bibliothèque ʿAbd al-Rahman al-Hadhrami ont été rédigés sur un papier filigrané arborant un croissant à profil humain avec, à droite et à gauche de sa partie supérieure, deux étoiles à six branches, insérés dans un écu à double contour. Il porte, sur une ligne, la contremarque en langue et caractères arabes : « *Waraq Abū Šibbāk al-aṣīlī* (ورق ابو شِباك الأصيلي) »[41] [ill. 4.4a, b].

Le filigrane et la contremarque apparaissent sur un même bifeuillet ; les dimensions des feuillets sont indiquées pour chaque exemple. On compte 2,4 cm entre les bords internes de deux chaînettes et 2,7 cm entre les bords externes de 20 vergeures.

Le ms. m/ḥ 35 contient une copie manuscrite datée de 1358/1939–1940. Il est composé d'un total de 7 feuillets, rassemblés en un ternion + 1 feuillet. Les dimensions maximales d'un feuillet sont de 24,5 × 18 cm[42].

Trois des 13 codex de la collection (*maǧmūʿ*) 50 ont été rédigés sur ce même papier, les 50/6, s.d., 50/7, daté de 1356/1937–1938 au colophon, et le 50/10, daté au colophon de 1361/1942–1943. Les manuscrits 50/6 et 50/10 ont très certainement été copiés à Zabid, si l'on se reporte au nom des scribes.

---

40 L'analyse a été réalisée par l'INP-Pagora de Grenoble en novembre 2011, grâce au soutien financier de l'UMR 7192 (Cnrs).

41 Les lettres sont ponctuées et la vocalisation présente a été reportée. Il n'y a pas de signe de redoublement (*šadda*) sur le *bāʾ*.

42 *CCBMZ*, I. 2, 2008.

ILLUSTRATION 4.4A *Type C. Filigrane. Ms. UbL Or. 25.744/3, daté 1356/1937*

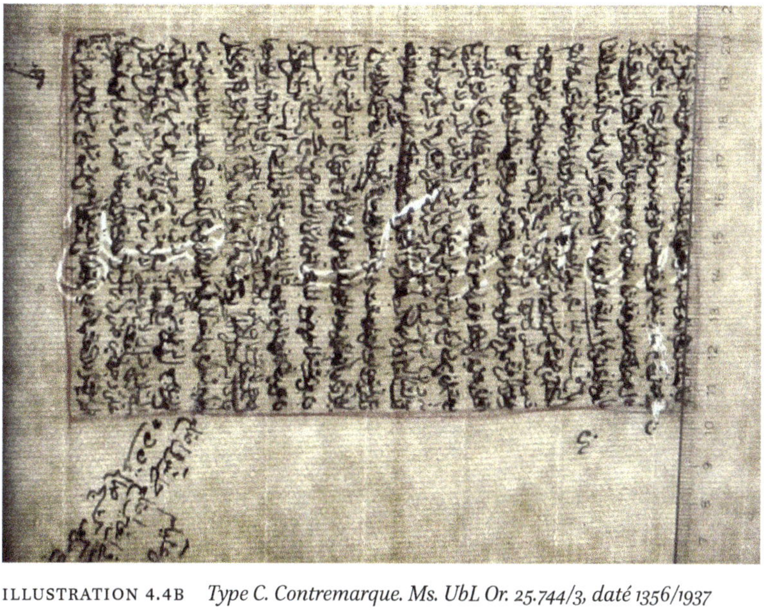

ILLUSTRATION 4.4B *Type C. Contremarque. Ms. UbL Or. 25.744/3, daté 1356/1937*

Le codex 50/6 est formé d'un quaternion, le codex 50/7, de 2 quinions + 1 feuillet, enfin, le codex 50/10, d'un quinion + 1 ternion + 2 feuillets séparés. La dimension maximale d'un feuillet est la même, 24 × 18 cm[43].

On observera que dans aucun de ces codex ne se trouvent de papiers de type A, alors que l'on rencontre couramment des codex écrits sur différents papiers de la gamme Abū Šubbāk. C'est le cas des manuscrits suivants de l'UbL :

- UbL 25.744/3, dont les cahiers sont en majorité des quaternions, 24,4 × 17,8 cm ; le copiste a utilisé concurremment du papier A.1. et C pour copier le même texte ;
- UbL 23.447, sur 1 binion, 24,2 × 18,5 cm, suivi d'un second texte sur papier A.1., à l'exception du bifeuillet central du dernier cahier qui est un *Waraq Abū Šibbāk* ;
- UbL 23.971, dont tous les cahiers n'ont pu être décomptés, mais qui comporte des quaternions et binions, 12,3 × 9 cm, a un cahier composite, fait de papiers *Waraq Abū Šibbāk* et A.2.

Dans le fonds de la Fondation Hā'il Saʿīd, à Taez, le ms. 51 est rédigé sur un papier de type C. Les informations concernant les vergeures et chaînettes n'ont pu être relevées. Le manuscrit n'est pas daté au colophon.

Enfin, le papier « *abū šibbāk* » noté par T. Walz est mentionné pour le Soudan de l'Est de la fin du xixe s., mais on ne dispose d'aucune autre indication.

Les dimensions des différentes marques du papier sont les suivantes (corpus Zabid et UbL)[44] :

a. entre 2 fils de chaînette : 2,4 à 2,5 cm ;
b. pour 20 vergeurs : 2,8 à 2,9 cm ;
c. H maximale de l'écu (entre les pointes supérieure médiane et inférieure) : 7,4 à 8 cm ;
d. distance entre les deux pointes supérieures à dextre et sénestre de l'écu : 4,3 à 4,5 cm ;

---

43 *CCBMZ*, I. 2, 2008.

44 Ces dimensions ont été dégagées sur très peu d'occurrences. Le relevé d'un feuillet du ms. m/ḥ 50/7 montre une hauteur atypique de l'écu, atteignant 8,3 cm, tandis que sur un autre feuillet, la hauteur est de 7,5 cm, dans la moyenne donc.

e. l maximale de l'écu (partie renflée) : 4,9 à 5 cm ;
f. L maximale du texte de la contremarque : 16,5 cm.

Au total, le papier de type C a pour le Yémen, un spectre court d'utilisation, qui s'étend de 1356/1937 à 1363/1944. Il n'a pas été relevé en Éthiopie.

Type D. Papier **Waraq Abū Šubbāk/Andrea Galvani Pordenone**

Il s'agit d'un papier filigrané, arborant un croissant à profil humain, inséré dans un écu à double contour, avec deux contremarques :

du côté droit, sur 2 lignes, en langue italienne et caractères latins :

[1re ligne] *Andrea Galvani*
[2e ligne] *Pordenone* ;

du côté gauche, sur une ligne, en langue et caractères arabes :

Waraq Abū Šubbāk (وَرَقْ ابُو شِبَاكْ)[45].

Le profil humain du croissant est orienté vers la gauche. Les deux contremarques figurent dans le champ du croissant, le texte arabe étant situé à hauteur de l'intervalle entre les deux lignes du texte italien. L'écart interne entre deux fils de chaînette est de 2,5 cm et la distance mesurée à partir des bords externes de 20 vergeures, de 3,1 cm.

Ont été trouvées à ce jour seulement quatre occurrences de ce papier, toutes dans des manuscrits d'Éthiopie. Deux dans des manuscrits consultés à l'IES, à Addis Abeba, les mss. IES 4574/1 et 4627, et deux dans un manuscrit de la collection Weiner, les W59 et W74. D'après les informations textuelles, le ms. IES 4627 a de très fortes chances d'avoir été copié en Éthiopie. Le ms. W74 peut être aussi éthiopien du fait du texte qu'il contient et d'éléments physiques ; c'est peut-être le cas encore du ms. W59.

Le ms. 4574 est une collection formée de deux codex. Le premier d'entre eux, ms. 4574/1, constitué d'un seul quaternion, contient un traité de logique sans titre et auteur, ni colophon ; de très nombreux commentaires figurent dans les marges. C'est celui qui a été rédigé sur le papier de type D [ill. 4.5.a, b. ; les deux feuillets ne coïncident pas bord à bord, parce que le papier s'est légèrement recourbé]. La dimension maximale des feuillets est de 24,5 × 18,5 cm. Il est difficile de dénombrer les cahiers du second codex, plus volumineux.

45 Les lettres sont ponctuées et la vocalisation présente a été reportée.

ILLUSTRATION 4.5A *Type D.* Waraq Abū Šubbāk/Andrea Galvani Pordenone. *Ms.* IES *4574* (*18*). *Addis Abeba, échelle 1*
© AIMABLE AUTORISATION IES. PHOTOGRAPHIE ANNE REGOURD, INFOGRAPHIE HÉLÈNE DAVID-CUNY

ILLUSTRATION 4.5B *Gros plan, bord. Ms.* IES *4574* (*18*). *Addis Abeba, échelle 1*
© AIMABLE AUTORISATION IES. PHOTOGRAPHIE ANNE REGOURD, INFOGRAPHIE HÉLÈNE DAVID-CUNY

Le texte comporte un colophon, qui indique avec précision le nom du copiste, Muḥammad b. ʿArab al-Artaǧī *nāhiyatān* al-Aykbarī *qariyatān al-šāfiʿī maḏhabān al-ašʿarī iʿtiqādān al-qādirī ṭarīqān*, et la date de copie, ǧumādā al-ṯānī 1355/1936[46]. Ce second codex a été copié sur du papier Abū Šubbāk de type A (sans doute A.2., du fait des dimensions de l'écu et de la couleur actuelle du papier). La dimension maximale des feuillets est de 24 × 18 cm. La proximité des deux « Abū Šubbāk » a permis de les comparer. Ils n'ont pas évolué de la même manière : le type D tire vers le blanc, tandis que le type A a bruni.

Le ms. 4627 n'a pas de colophon. Il a été estimé du xixe s.[47]. Pour des raisons de conservation, il a été impossible de dénombrer les cahiers. La dimension

46 A. Gori, *A Handlist of the Manuscripts in the Institute for Ethiopian Studies*, 2014, IES 4574, 64.

47 *Ibid.*, 70–71.

maximale d'un feuillet est de 24,5 × 18 cm. On notera que les bifeuillets se sont souvent désolidarisés, le long de la pliure centrale.

Le ms. Weiner W74 n'est pas daté, mais on note la présence d'un autre papier, fait à la machine[48].

En W59/EMIP 00563, l'autre manuscrit Weiner, l'année, donnée seule au colophon (avant-dernier feuillet), a été effacée. Autour, figurent des tentatives de restauration de la date : au-dessous, on a 1363/1943–1944 (en chiffres et en lettres), puis au-dessous encore, « sana 1313(/1895–1896) hiǧriyya nabawiyya », et, à 360°, « bi-ism Allāh 1154 (?) 313 ». De toutes ces dates certainement suggérées par des lecteurs, c'est celle de 1363 qui semble le mieux correspondre à ce qui reste visible de la date inscrite au colophon.

La contremarque du papier D mentionne la maison Galvani, par le biais de l'un de ses fils, Andrea, qui a vécu durant la première moitié du XXe s. (1797–1855)[49]. Cette contremarque, qui mentionne aussi le moulin de Pordenone, a continué d'être en usage bien après la mort d'Andrea Galvani, comme l'attestent des manuscrits de Zabid, produits entre 1291/1874–1875 (m/ḥ 14, copié dans le ribāṭ de Yaḥyā b. ʿUmar al-Ahdal à Zabid) et 1337/1918–1919, le ms. UbL Or. 6980, et un manuscrit du Nigéria écrit en 1924[50]. C'est une autre caractéristique du papier D qui permet d'avancer dans sa datation : il est produit au cylindre [ill. 4.5.b.]. Les premiers signes d'une mécanisation de la production de papier dans les moulins Galvani se font jour à partir de 1889 ; le processus de modernisation des moulins se poursuit ensuite progressivement[51].

48 Le filigrane, qui porte un texte et un motif à l'intérieur d'un cercle, est malheureusement partiel.

49 Voir G. Ganzer, *Andrea Galvani, 1797–1855 : cultura e industria nell'Ottocento a Pordenone*, 1994, en particulier l'arbre généalogique.

50 Voir également pour le Yémen : M.-G. Guesdon, « Manuscrits de provenance yéménite donnés à la Bibliothèque nationale par Pierre Bardey en 1930 », *Chroniques yéménites* 13 (2005), 67 et 68 ; Égypte : T. Walz, « The Paper Trade of Egypt and the Sudan in the Eighteenth and Nineteenth Centuries », 1985, 29sq. ; Soudan : T. Walz, *ibid.*, 39–41 ; A. Brockett, « Aspects of the physical transmission of the Qur'ān », 1987. Le ms. UbL Or. 6980, A. Regourd, « Le ms. Or. 6980 de la Bibliothèque universitaire de Leyde. I. Premiers éléments de datation et de localisation du manuscrit : apport d'un papier Andrea Galvani inédit », juillet 2015. Pour le Nigéria, M. L. Biddle, « New strategies in using watermarks to date sub-Saharan Islamic manuscripts », 2017, papier AGmm6.

51 Le processus d'industrialisation a lieu aux alentours de 1883–1895, voir I. Mattozzi, « I Galvani, fabbricanti di carta (1744–1855). Un modello di formazione dell'imprenditorialità? », 1994, p. 36 ; M. L. Biddle, « New strategies », 2017, chap. « Galvani Countermarks and Watermarks in the Machine Era ».

Ce n'est pas le premier exemple de contremarque à langues mixtes, le présent volume l'atteste pleinement. Dès 1985, pour l'Afrique, Walz en relève deux au Nigéria, l'un en caractères latins et arabes, l'autre en *'ağamī*[52]. Walz note que ce dernier exemple corrobore la production d'un papier pour la satisfaction d'un marché donné.

Pour le papier de type D, on peut poser une mise en production potentielle à partir de 1889.

Postérité et longévité du papier Abū Šubbāk

Type E. Papiers Abū Šubbāk du souk

Sur l'aire géographique que l'on vient de parcourir, le papier Abū Šubbāk est présent sur une période qui couvre approximativement la première moitié du xxᵉ s. L'appellation de papier « Abū Šubbāk », c'est-à-dire de papier « à la fenêtre », pour un article de qualité, à filigrane composé d'un croissant de lune à profil humain dans un écu, est attestée dès 1303/1886[53]. La qualité Abū Šubbāk est restée, elle, dans les mémoires yéménites, jusqu'à nos jours comme un synonyme d'excellent papier, robuste et pérenne, préconisé dans les cas où l'on veut coucher par écrit des actes légaux.

Sa robustesse explique aussi son utilisation fréquente comme feuillet de garde des manuscrits[54]. Dans les années 2000, nous avons pu acheter au souk de Zabid, « un Abū Šubbāk », soit un feuillet complet de papier[55]. Il s'agit d'un

52 Voir T. Walz, « The Paper Trade of Egypt and the Sudan in the Eighteenth and Nineteenth Centuries », 1985, 42, repris par J. M. Bloom, « Paper in Sudanic Africa », 2008, 54, et M. L. Biddle dans ce volume.

53 Voir l'article de Jan Just Witkam dans ce volume. Différents papetiers italiens ont eu recours au filigrane à écu à profil humain.

54 Nous en avons personnellement constaté de nombreux exemples que nous n'avons pas toujours notés ; voici quelques références : Bibliothèque 'Abd al-Rahman al-Hadhrami, m/ḥ 7, m/ḥ 50 ; Sanaa, Dār al-maḫṭūṭāt, ms. 2774, 'ulūm ḫafiyya 22, M. S. al-Malīh & A. M. 'Aysawī, *Fihris maḫṭūṭāt al-Maktaba al-ġarbiyya bi-al-Ğāmiʿ al-kabīr bi-Ṣanʿā'*, [1978], 432, décrit dans A. Regourd, « Images de djinns et exorcisme dans le *Mandal al-sulaymani* » (avec éd. de texte et trad.), 2011, 254.

55 L'évolution du nom propre d'une marque à un générique s'observe pour d'autres produits : on dit couramment en français « un frigidaire » pour un réfrigérateur – Frigidaire étant une marque –, ou, en anglais, « a hoover » pour un aspirateur (*vacuum cleaner*), qu'il soit ou non de marque Hoover.

ILLUSTRATION 4.6 *Type E. Souk de Zabid, années 2000*
© PHOTOGRAPHIE ANNE REGOURD

papier non filigrané, de couleur jaune [ill. 4.6]. Les dimensions d'un feuillet original sont de 50 × 35 cm. Les chaînettes sont perpendiculaires à la feuille, envisagée dans sa longueur. La distance entre les bords internes de deux chaînettes est de 2,3 cm. L'espacement couvert par 20 vergeurs, mesuré à partir des bords externes, représente entre 3,2 et 3,3 cm. Il ne comporte ni filigrane, ni contremarque.

Statistiques

Seuls les textes manuscrits datés au colophon ont été retenus. Ils proviennent des bibliothèques et collections suivantes :

BnF	Bibliothèque nationale de France, Paris, France
FHS	Fondation Hā'il Saʿīd, Taez, Yémen
IES	Institute for Ethiopian Studies, University of Addis Ababa, Éthiopie
ʿArHZ m/ḥ	Bibliothèque privée ʿAbd al-Raḥmān al-Ḥaḍramī, Zabid, Yémen
UbL	Universiteitsbibliotheek Leiden, Leyde, Pays-Bas
W	Collection privée Gerald & Barbara Weiner, États-Unis

Cote	Type A Abū Šubbak A.1.	Type A Abū Šubbak A.2.	B	Type C *Waraq Abū Šibbāk*	Type D Galvani/ Abū Šubbāk
		Yémen			
ʿArHZ m/ḥ 60		1263/1846–1847			
UbL Or. 25.288/3		ṣafar 1324/ 03–04.1906			
UbL Or. 23.256		muḥarram al-ḥaram 1328/1910			
UbL Or. 6362/2		3 (?) šaʿbān 1328/10.08.1910			
ʿArHZ m/ḥ 66/1		22 ǧumād aḫar 1328/ 01.07.1910–1911			
FHS 107		1328/1910–1911			
UbL Or. 25.174/5		24 rabiʿ awwal 1329/25.03.1911			
UbL Or. 25.174/2		28 ǧumādā aḫir 1329/26.06.1911			
ʿArHZ m/ḥ 50/9		1330/1911–1912			
UbL Or. 25.174/3		ǧumādā al-awwal & al-aḫir 1330/04–06.1912			
UbL Or. 25.287/3		šaʿbān 1332/06–07.1914			
UbL Or. 25.174/7		7 (?) rabīʿ awwal 1333/23.01.1915			
UbL Or. 25.735		9 rabīʿ al-awwal 1334/15.01.1916			
UbL Or. 25.736		différentes dates de l'année 1336/1918			
UbL Or. 23.268		rabīʿ ṯānī 1336/01–02.1918			
UbL Or. 25.283/4		6 ǧumādā aḫir 1336/19.03.1918			
UbL Or. 25.283/6		salḫ ǧumādā aḫir 1336/04.1918			

Cote	Type A Abū Šubbak A.1.	Type A Abū Šubbak A.2.	B	Type C *Waraq Abū Šibbāk*	Type D Galvani/ Abū Šubbāk
UbL Or. 25.283/7		8 rağab 1336/19.04.1918			
UbL Or. 25.283/3		10 rağab 1336/21.04.1918			
UbL Or. 25.283/8		rabiʿ al-aḫir 1337/01–02.1919			
UbL Or. 23.449		23 ğumādā awwal 1338/13.02.1920			
UbL Or. 25.663/3		ğumād aḫir 1338/27.02.1920			
UbL Or. 25.662		24 rağab 1338/13.04.1920			
UbL Or. 6368		8 ṣafar 1341/30.09.1922			
UbL Or. 6353a		1342/1923–1924			
UbL Or. 6353b		1342/1923–1924			
UbL Or. 25.175/1	17 ṣafar 1345 /27.08.1926				
UbL Or. 25.175/2	17 rabīʿ al-awwal 1345/25.09.1926				
UbL Or. 6330	10 muḥarram 1346/10.07.1927				
UbL Or. 23.407/1–4	1346/1927–1928				
UbL Or. 23.972	22 al-ḥiğğa 1346/11.06.1928				
UbL Or. 6634	24 ḏū al-ḥiğğa 1346/13.06.1928	24 ḏū al-ḥiğğa 1346/13.06.1928			
UbL Or. 6643/ 1–3		Entre 10.1929 et 08.1930			
UbL Or. 6360	21 ğumādā al-aḫir 1349/13.11. 1930				
UbL Or. 6357a	5 ğumādā al-aḫir 1350/18.10.1931				
UbL Or. 25.663/1	10 šaʿabān 1350/ 21.12.1931				

TABLE (suite)

Cote	Type A Abū Šubbak A.1.	Type A Abū Šubbak A.2.	B	Type C *Waraq Abū Šibbāk*	Type D Galvani/ Abū Šubbāk
UbL Or. 23.448	5 safar 1351/ 10.06.1932	5 ṣafar 1351/10.06.1932			
ʿArHZ m/ḥ 50/8		1351/1932–1933			
ʿArHZ m/ḥ 50/11	1351/1932–1933				
UbL Or. 6637	šawwāl 1351/ janv.–fév. 1933				
UbL Or. 25.745	28 ǧumād aḫir 1355/ 15 sept. 1936				
ʿArHZ m/ḥ 50/12	1355/1936–1937				
ʿArHZ m/ḥ 50/12		1355/1936–1937			
UbL Or. 25.744/3	raǧab 1356/ sept.–oct. 1937			raǧab 1356/09–10.1937	
UbL Or. 23.975[a]	muḥarram 1356/ mars–avril 1937				
ʿArHZ m/ḥ 50/4	1356/1937–1938				
ʿArHZ m/ḥ 50/7				1356/1937–1938	
UbL Or. 23.447	rabīʿ awwal 1357/05.1938				
ʿArHZ m/ḥ 50/3	1358/1939–1940	1358/1939–1940			
ʿArHZ m/ḥ 35				1358/1939–1940	
ʿArHZ m/ḥ 50/10				1361/1942–1943	
UbL Or. 23.971		13 rabīʿ awwal 1363/08.03.1944		13 rabīʿ awwal 1363/08.03.1944	
UbL Or. 23.407/8	8 ṣafar 1372/ 28.10.1952				

Cote	Type A Abū Šubbak A.1.	Type A Abū Šubbak A.2.	B	Type C *Waraq Abū Šibbāk*	Type D Galvani/ Abū Šubbāk
UbL Or. 23.407/10	15 ṣafar 1372/ 4.11.1952				
UbL Or. 23.407/6, 7	11 rabīʿ awwal 1372/ 29.11.1952				
UbL Or. 23.407/5	1372/1952–1953				
UbL Or. 23.407 /11	3 ǧumādā awwal 1372/19.01.1953				
ʿArHZ m/ḥ 1[b]	1375/1956				
UbL Or. 25.287/4		šawwāl 1386/01–02.1967			
TOTAL	25	34	1	6	0 66
		Éthiopie			
W41/EMIP 00543		1327/1908–1909			
W28/EMIP 00530		1328/1909–1910			
IES 4624	rabīʿ al-awwal 1328/ 03–04.1910				
W70/EMIP 00574		1329/1910–1911			
IES 5505		2[e] moitié 1329/07.1911			
W39/EMIP 00541		1331/1912–1913			
W66/EMIP 00570		1332/1913–1914			
IES 5517/1		19 rabīʿ al-ṯānī 1337/22.01.1919			
IES 2637	14 ǧumādā al-awwal 1342/ 23.12.1923				

TABLE (cont.)

Cote	Type A Abū Šubbak A.1.	Type A Abū Šubbak A.2.	B	Type C *Waraq Abū Šibbāk*	Type D Galvani/ Abū Šubbāk
BnF Ar. 7264/1		1345/1926–1927			
BnF Ar. 7264/2		dū al-qaʿada 1345/ 05.1927[d]			
IES 5509		01 ramaḍān 1348/31.01.1930			
W42/EMIP 00544	rabiʿ I 1353/1934–1935				
IES 4574/2	ǧumādā al-ṯānī 1355/ 08–09. 1936				
W59/EMIP 00563	1313/1895–1896— 1363/1943–1944				1313/ 1895–1896— 1363/ 1943–1944[e]
TOTAL	9	6	0	0	1 = 16[f]

a Il s'agit des feuillets de garde inférieur et supérieur, dont l'un porte la date mentionnée.

b Reproduit dans le CCBMZ, I. 1. *Les papiers filigranés*, 2008, clichés sous n° 003.

c Le ms. UbL Or. 25.179 a été écrit sur du papier de type A.1., au moins jusqu'au f. 75, mais l'année portée au colophon, f. 192r, dimanche 11 ṣafar 1031/26 déc. 1621, est douteuse.

d Voir la note 15.

e Au sujet de l'incertitude sur la date au colophon du W59/EMIP 00563, voir Type D.

f Le ms. W46/EMIP00550, écrit sur du papier de type A.1., n'a pas été compté. L'année est apparemment libellée en lettres, « sana ġayn bāʾ šīn », ce qui donne : 8 ḏū al-ḥiǧǧa 1302/ 09–10.1885, à côté de 3 chiffres, 135, suivis de la lettre *hāʾ*, peut-être pour [1]135 *h*[*iǧra*] … ? Il paraît invraisemblable que la seconde date soit celle de la copie. Mais un doute subsiste au sujet de la première, pourtant très intéressante pour notre propos.

De la circulation du papier à quelques réflexions sur le commerce : une « zone » « *Abū Šubbak* » / « *Abū Šibbāk* »/ « *Abū Šubbāk* » ?

Aucun papier Abū Šubbāk n'a été signalé en Égypte par Walz, qui s'est appuyé à la fois sur l'examen de documents d'archives et de codex[56]. Cependant, un doute subsiste car il fait courir son étude jusqu'en 1880, une date charnière à la fois dans les rapports de force de la région Égypte-Soudan qui sont le cadre de son étude et pour l'apparition des papiers Abū Šubbāk – un doute que les deux autres publications sur des filigranes de documents du Caire allant jusqu'à la fin du xixe s. ne peuvent chasser. Walz note cependant à partir de Shuqayr un « *abu shibbāk* » pour le Soudan de l'Est, à la fin du xixe s., qui échappe au spectre d'utilisation que nous avons dégagé grâce à une recherche sur les manuscrits de Zabid, de l'IES et de Leyde qui incluait pourtant les codex datés du xixe s. L'appellation de « papier Abū Šubbāk », ou « celui à la fenêtre », est attestée dès 1886 à La Mecque : reste à identifier des papiers correspondant à ceux de notre corpus dans les manuscrits de La Mecque, Médine ou Djedda[57]. Une vérification *a contrario* montre qu'aucun des 24 manuscrits sur les 50 du fonds Archinard (Bibliothèque nationale de France) provenant d'Afrique de l'Ouest, qui ont été écrits sur papiers filigranés sur une période allant du xviiie à la première moitié du xxe s., ne correspond au papier étudié. Il faudrait poursuivre l'enquête *a contrario* sur des manuscrits produits sur le pourtour méditerranéen. Au Nigéria, les papiers filigranés à marques à écriture arabe sont autres[58].

Il est vraisemblable que quelques-uns des manuscrits d'Éthiopie étudiés aient été copiés sur place. Pourtant, il ne faut pas perdre de vue les contacts récurrents des lettrés de ce pays avec ceux du Yémen[59]. Des familles de Sayyid-s, donc de lettrés, du Hadramaout, ont émigré vers l'Indonésie et la Malaisie, au

---

56 T. Walz, « *The Paper Trade of Egypt and the Sudan in the Eighteenth and Nineteenth Centuries* », 1985, 29 et 43, notes 4 et 6, où il indique qu'il a consulté, au Caire, 276 vols. d'archives du Maṣlaḥat al-Šarh al-ʿAqārī, ainsi que 60 manuscrits de la Bibliothèque nationale, Dār al-kutub ; *ibid.*, 2011, 73–74 et notes 8, 76, et 12, 77–78.

57 Voir l'article de J. J. Witkam dans ce volume.

58 Voir T. Walz, « The Paper Trade of Egypt and the Sudan in the Eighteenth and Nineteenth Centuries », 1985, 41–42 ; *ibid.*, 2011, 103–104 ; M. L. Biddle sur les papiers du Nigéria dans ce volume.

59 Voir, par ex., la biographie représentative du Šayḫ al-Asad qui, originaire d'al-Ūsī en Éthiopie, est devenu *ʿālim al-ʿallāma* et s'est établi à Zabid, dans : « Šayḫ ʿUlamāʾ Zabīd— *al-šayḫ al-ʿallāma* Asad Ḥamza b. ʿAbd al-Qādir al-Ūsī », *Mağallat al-Ğadwa* 7, 1425/2004, 99–103.

moins depuis le xviie s.[60]. Entre le Yémen et l'Éthiopie, des manuscrits ont circulé, mais, c'est possible aussi, du papier. La reconstruction de l'histoire des bibliothèques et des manuscrits, ainsi qu'une base statistique plus importante traduisant la présence de ce papier dans des codex éthiopiens devraient nous en apprendre plus. Pour l'Indonésie, on ne peut écarter la circulation de manuscrits en provenance du Hadramaout[61]. La même question se pose pour les papiers des manuscrits de la Mosquée al-Riyadha à Lamu (Kenya), fondée par Ṣāliḥ b. ʿAlawī Ǧamāl al-Layl (1853–1936), car ces derniers indiquent des liens avec le Yémen[62]. Pour l'instant, dans cet ensemble, le foyer le plus probable d'importation des papiers de type A-B est le Yémen.

Quant au type D, quatre occurrences seulement ont été repérées, exclusivement en Éthiopie, dans des manuscrits dont un seul est pourvu d'un colophon. Celui-ci a la particularité de donner plusieurs dates, ouvrant à une fourchette de datation d'une cinquantaine d'année : elle ne bouleverse cependant pas le tableau général.

Production et commerce : vers une compétition entre plusieurs fabricants ?

Nous avons dégagé deux groupes de papier Abū Šubbāk, les *Beyāḍ Abū Šubbak* et les *Waraq Abū Šibbāk/Abū Šubbāk*.

Le premier groupe recouvre les types A et le papier B. Les papiers A les plus anciens ont été produits à la machine au cylindre. Le type A.2. a servi à fabriquer des codex de 1906 à 1967, soit un spectre d'utilisation d'une soixantaine d'années ; c'est aussi le cas du type A.1., pour une période allant de 1908–1909 à 1956, soit un peu moins d'une cinquantaine d'années. Le papier B est signalé en 1942–1943, il s'agit d'un papier produit à la machine au cylindre, utilisé dans l'imprimerie.

---

60 Voir note 26. Cet ouvrage collectif contient d'autres articles, plus monographiques, qui donnent des généalogies remontant à des familles de Sayyids hadramis.

61 Voir même note.

62 Voir A. K. Bang, « The Riyadha Mosque Manuscript Collection in Lamu: A Ḥaḍramī Tradition in Kenya », 2014, 125–153. Le ms. Lamu RM 44 a circulé du Hadramaout ou de l'océan Indien en direction de Djakarta, puis est retourné à l'Est de l'Afrique pour parvenir enfin à Lamu entre 1253/1837–1838 et la deuxième moitié du xxe s. Le catalogage de la collection n'a fait apparaître jusqu'ici que des papiers fabriqués en Europe et aucun papier Abū Šubbāk.

Le lieu de production de ces papiers reste à identifier. Il est difficile de se prononcer sur la langue du texte de la contremarque, turc-ottoman ou arabe, de regrouper ces papiers par ce biais. Le mot شبّك comporte un *alif* en arabe ; en *osmanlı*, il renvoie aux grilles des mausolées, mais les fenêtres orientales sont des chassis à grillage. La mention à Istanbul est donnée par la *nisba isṭambūlī*, « d'Istanbul », qui vient avant *ʿālī aṣīlī* deux mots qualifiant la qualité du papier : au-delà d'une provenance, il pourrait s'agir d'une référence à un excellent papier ou à un format. La seconde occupation ottomane du Yémen prend fin en 1918. Une fabrication turque des papiers de type A-B reste à prouver. Force est de constater – sur la base d'un nombre d'étude encore restreint il est vrai – que le secteur de la papèterie en Turquie, bien que mécanisé dès la première moitié du xixᵉ s., connaît des hauts et des bas : les firmes privées naissent et s'éteignent, elles ne durent pas[63]. Si l'on continue de chercher dans cette direction, vers où se tourner pour trouver des papiers exportés produits sur le sol turc ? On pense inévitablement à l'administration ottomane, grosse consommatrice de papier. Un premier acompte a été fourni par Nikolaev, Velkov et Andreev, qui ont publié quatre volumes de papiers filigranés provenant des archives ottomanes conservées à la Bibliothèque nationale de Bulgarie, à Sofia[64]. La présence de papiers d'origine ottomane, concurremment avec des

63 Pour la période intéressant cette enquête, n'ont été répertoriées sur le sol turc à ce jour que les papèteries suivantes :

- la papèterie Hamidiye de Beykoz, du nom de la rivière l'alimentant, située sur le Bosphore. Elle naît à la toute fin du xixᵉ s., vers 1890. L'imprimerie, fondée à l'initiative d'Osman Bey, chambellan du Sultan Hamid ii, est de grande taille et puissamment équipée : on note en particulier 4 machines à papier anglaises Masson Scott & Co., un vaste atelier d'apprêt et de façonnage du papier, ainsi qu'une centrale thermique. La paille et le chiffon sont ses matières premières. Sa disparition, vers 1915, serait due à des intrigues politico-économiques.
- l'Anatolie, bastion de l'industrie papetière moderne, détient la fabrication de la cellulose et du papier. Dans le plan quinquennal de 1933–1938, était prévue la création d'une grande papèterie. Voir A. Regourd, « Paper Trade on the Red Sea (XIXᵉ c.–first half of the XXᵉ c.) », « Overview of paper-making in 19th & 20th c. Turkey », à paraître.

Les composantes de la pâte de B ne sont en revanche pas incompatibles avec une production sur le sol turc, voir M. A. Kâğıtçı, *Historique de l'industrie papetière en Turquie*, 1976, 34–35.

64 V. Nikolaev, *Watermarks of the Mediaeval Ottoman Documents in Bulgarian Libraries, Watermarks of the Ottoman Empire*, vol. 1, 1954 ; A. Velkov & S. Andreev, *Filigranes dans les documents ottomans*. Vol. 1 : *Trois croissants*, 1983 ; A. Velkov, *Les filigranes dans les documents ottomans. Divers types d'images*, 2005 ; S. Andreev, *Les filigranes dans les documents ottomans. Couronne*, 2007.

papiers d'autres origines, y est attestée pour la période du xvie au xviiie s.[65]. Malheureusement pour notre étude, la sélection des deux auteurs s'arrête à des documents d'archive datant de 1820[66].

Les papiers Abū Šubbāk ne sont pas davantage enregistrés dans le *Phillips Paper Trade Directory of the World* (années 1906/1907–1946/1947)[67].

Le second groupe, recouvrant les types C et D, a une contremarque en arabe et ne fait plus allustion à Istanbul. Celle du type D, qui fait référence à la maison Galvani, permet d'envisager un papier en production à partir de 1889. L'information donnée par Walz citant Shuqayr, qui mentionne un papier appelé « *abū šibbāk* » parmi les importations européennes vers le Soudan à la fin du xixe s., fait donc sens. Mais rapporter ce papier à ceux de notre corpus est pour l'instant compliqué par l'absence de toute description à notre disposition. En arabe dialectal, « *šibbāk* » se dit pour « fenêtre », gardant possible un type C ou D. Sur notre corpus à ce jour, le type D a été exclusivement recueilli en Éthiopie et le spectre d'utilisation du type C, présent, lui, uniquement au Yémen et plus tardif, s'étend de 1937 à 1944. Le rapport de Shuqayr dont Walz se fait l'écho suggérant un approvisionnement indépendant, rend une enquête sur les codex soudanais à partir de la seconde moitié du xixe s. d'autant plus pertinente, enquête qu'il faudrait compléter en Égypte. On peut au total affirmer que le papier D était fabriqué dans le Frioul, mais le fabricant de C reste à préciser.

Conclusions

Le premier objectif de ce travail débuté en 2010 était de mettre en évidence l'existence de plusieurs papiers Abū Šubbāk. L'identification et le classement des différents exemples connus à ce jour ont fourni un spectre d'utilisation de ces papiers dans l'espace et le temps.

Au cours de l'enquête, il est toutefois apparu que les papiers Abū Šubbāk ne devaient pas retenir notre attention par le seul jeu de la classification des papiers entre papiers à marques à caractères latins et papiers à marques à caractères non-latins, apparus tardivement, mais par leur emploi omniprésent et régulier dans la première moitié du xxe s. : il s'agit en effet de papiers courants,

65 A. Velkov & S. Andreev, *Filigranes dans les documents ottomans*. Vol. 1 : *Trois croissants*, 1983, 14.

66 Communication personnelle d'A. Velkov, 2013. Dans l'introduction de leurs ouvrages respectifs, Velkov et Andreev se sont toujours reportés à une période allant du xvie au xviiie s.

67 *Phillips Paper Trade Directory of the World*, 1906/1907–1946/1947.

à usages multiples (codex et documents manuscrits, feuillet de garde, imprimerie), au Yémen. Ces indications laissent supposer un approvisionnement sur la durée.

Se référer à « Abū Šubbāk » dans les marques du papier répond à un marché précis de locuteurs arabes ou sensibilisés à cette langue et de connaisseurs de papiers de ce nom : en écho, l'enquête montre la présence de ces papiers dans des pays du pourtour de la mer Rouge ; en Éthiopie, le papier est un marqueur des manuscrits en arabe ou en arabe *ʿağamī*[68].

Au Yémen, la production de codex montre qu'il a pénétré dans le Nord, jusque dans la région de Ṣaʿda, aussi bien qu'à Zabid, dans la plaine côtière de la mer Rouge. Le papier B était présent à Sanaa, servant aux presses de l'Imam. Pour la « période Abū Šubbāk », Aden est sous contrôle anglais, tandis que l'Imam Yaḥyā b. Muḥammad b. Yaḥyā Ḥamīd al-dīn (r. 1918–1948), depuis Sanaa, reprend Taez et des points-clés de la Tihāma, dont des ports et Zabid. L'Imam pratique une politique isolationiste, mais doit faire face à une opposition libérale, active à Aden, où presses et papiers colportent ces idées[69]. En Éthiopie, la production de codex montre que notre papier parvient à l'Ouest, dans le Harar, mais aussi dans le Centre-Nord du pays, au nord d'Addis Abeba, dans le Qāllu-Wällo, dans le Sud-Est du Wällo et dans le Wällo-Nord du Šäwa, tous lieux à forte proportion musulmane.

Un fabricant, italien, a été identifié. Après avoir créé un papier de bonne qualité, apprécié des consommateurs, qui le désignent du nom de « Abū Šubbāk » avant 1886, la maison Galvani, largement maître des marchés nigérien, égyptien et certainement soudanais au xixe s.[70], présente au Yémen, doit faire face à la concurrence, que décuple l'ère industrielle. Dès 1868, leurs ventes s'affaissent devant la production de papier machine[71]. À partir de 1889, les Galvani réagissent, commencent à équiper leurs moulins, aiguisent leur stratégie commerciale en « récupérant » à une date qui reste à déterminer la dénomination de « Abū Šubbāk » dans le texte de leur contremarque (type D) pour des papiers livrés vraisemblablement en Éthiopie, peut-être au Soudan. Leurs papiers à la machine du Nigéria au début du xxe s. montrent une

68 A. Regourd, « Papers », 2014, xlix.

69 Les exemples les plus célèbres sont le programme de réformes demandées par les libéraux, *Barnāmağ ḥizb al-Aḥrār al-yamanī bi-Ṣanʿāʾ*, à la date de 1357/1938–1939, imprimé sur les presses Fatāt al-Ğazīra à Aden, et l'organe libéral, d'Aden toujours, *Ṣawt al-Yaman*.

70 A. Brocket, « Aspects of the physical transmission of the Qur'ān in 19th-century Sudan: Script, decoration, binding and paper », 1987.

71 I. Mattozzi, « I Galvani, fabbricanti di carta (1744–1855). Un modello di formazione dell'imprenditorialità? », 1994, 37.

variation dans les contremarques, mais qui demeurent en alphabet latin. En Éthiopie et au Yémen, les cahiers d'écolier de Chine ou d'ailleurs servent de support aux livres manuscrits dès le début du xxᵉ s., suivis par des feuilles de papier-machine pré-lignées de différents formats, dont du « Foolscap » et, plus tard, du 21 × 29,7 cm, de plus ou moins bonne qualité, que l'on peut acheter au souk et plier pour réaliser un bifeuillet[72]. Entre 1906–1907 et 1946–1947, un rythme de production élevé est maintenu par les Galvani[73]. La perte de terrain et la mécanisation tardive de la firme l'ont donc poussée dans une logique de reconquête du marché, qui s'appuie sur l'allusion au papier « Abū Šubbāk ». C'est un chapitre encore inconnu de l'épopée de la très omniprésente maison Galvani qui surgit ici.

Le papier *abū šibbāk* soudanais, classé dans le second groupe (types C et D, notre chap. v), vient d'Occident. Il est pour l'instant difficile de se prononcer sur le(s) fabricant(s) et lieu(x) de fabrication des papiers A, présents, eux, des deux côtés de la mer Rouge. L'absence de tout texte en caractère latin et l'allusion à Istanbul/*isṭambūlī*, « d'Istanbul », font plutôt songer à une concurrence sur les marchés yéménite et éthiopien, au xxᵉ s., que se livraient plusieurs fabricants. La pâte des papiers A.2., suggérant une grande variété dans les matériaux utilisés, est une piste à poursuivre afin de lever un coin du voile qui s'étend sur le lieu de leur production.

L'autre chantier majeur qui reste à mener porte sur les papiers de manuscrits du Soudan, d'Égypte et de villes d'Arabie saoudite.

Bibliographie

Andreev, Stefan. 2007. *Les filigranes dans les documents ottomans. Couronne*, Sofia, éd. « Textes – Asparouh Trayanov ».

Anonyme. 1425/2004. « Šayḫ ʿUlamāʾ Zabīd – *al-šayḫ al-ʿallāma* Asad Ḥamza b. ʿAbd al-Qādir al-Ūsī », *Maǧallat al-Ǧadwa* 7, 99–103.

72 10,5 % des 172 manuscrits en 86 volumes de la Bibliothèque ʿAbd al-Raḥmān al-Ḥaḍramī, Zabid, colophons du milieu du xviiᵉ au milieu du xxᵉ s., est copié sur des cahiers d'écolier ; pour des exemples de papiers pré-lignés en Éthiopie, voir Regourd 2 dans ce volume.

73 « Between 1906/07 and 1946/7, with every issue of *Phillips Paper* Trade published, Galvani continued to report 'two machines 135 cm' and an annual production of 400 tons even though they closed the Porcia Rorai mill in 1929 and sold it in 1932 », M. L. Biddle, « New strategies », 2017.

Bang, Anne K. 2014. « The Riyadha Mosque Manuscript Collection in Lamu: A Ḥaḍramī Tradition in Kenya », dans : Anne Regourd (éd. invité), « Manuscrits en Transit : le cas du Yémen », *Journal of Islamic Manuscripts* 5/2–3, 125–153.

Biddle, Michaelle L. 2017. « New strategies in using watermarks to date sub-Saharan Islamic manuscripts », in: Andrea Brigaglia & Mario Nobili (éd.), *Arts and Crafts of Literacy in Sub-Saharan Africa*, Berlin, Walter de Gruyter.

Bloom, Jonathan M. 2008. « Paper in Sudanic Africa », chap. 4 dans : Shamil Jeppie, Souleymane Bachir Diagne (éd.), *The Meanings of Timbuktu*, Hsrc Press, Cape Town, Codesria/Dakar, 45–57 + 1 pl.

Brockett, Adrian. 1987. « Aspects of the physical transmission of the Qurʾān in 19th-century Sudan: Script, decoration, binding and paper », *Manuscripts of the Middle East* 2, 45–67.

Didot, Firmin. 1900. *Le Centenaire de la machine à papier*, Paris.

Dufour, Julien. Janvier 2013. « La *safīna* de Colin : une source importante pour l'histoire de la poésie *ḥumaynī* », *Chroniques du manuscrit au Yémen* 15.

Dufour, Julien & Regourd, Anne. À paraître. « Les *safīna*s yéménites », dans : Frédéric Bauden & Élise Franssen (éd.), *Authors as Scribes, Authors and Scribes. Autograph, Holograph and Authorial Manuscripts in Arabic Script. Proceedings of the International Conference held in Liège, 10–11 October*, 2013, E. J. Brill.

Ersoy, Osman. 1963. *XVIII. ve XIX. Yüzyıllarda türkiye'de kâğıt*, Ankara, Ankara Üniversitesi Basimevi.

Fihrist kutub al-ḫizāna al-mutawakkiliyya al-ʿāmira bi-al-ǧāmiʿ al-muqaddas bi-Sanʿāʾ al-muḥammiyya. 1312/1894–1895. Sanaa, Wizārat al-maʿārif al-mutawakkiliyya.

Freitag, Ursula, & Clarence-Smith, William G. 1997. *Hadhrami Traders, Scholars and Statesmen in the Indian Ocean, 1750s–1960s*, Leyde/New York/Cologne, E. J. Brill.

Freitag, Ursula. 2009. « Reflections on the Longevity of the Hadhrami diaspora in the Indian Ocean », dans : Ahmed Ibrahim Abushouk & Hassan Ahmed Ibrahim, *Hadhrami Diaspora in Southeast Asia. Identity Maintenance or Assimilation?*, Leiden/Boston, E. J. Brill, 17–32.

Ganzer, Gilberto. 1994. *Andrea Galvani, 1797–1855 : cultura e industria nell'Ottocento a Pordenone*, Pordenone, Studio Tesi.

Gori, Alessandro, coll. Anne Regourd, Jeremy R. Brown, Steve Delamarter. 2014. *A Handlist of the Manuscripts in the Institute for Ethiopian Studies*. Vol. 2 : *The Arabic Material of the Ethiopian Islamic Tradition*, Eugene (OR), Pickwick Publication, « Ethiopian Manuscripts, Texts, and Studies Series » 20.

Gori, Alessandro. 2015. « *Waqf* certificates of Qur'ans from Harar: A first assessment », dans : Alessandro Bausi, Alessandro Gori & Denis Nosnitsin (éd.), *Essays in Ethiopian Manuscript Studies. Proceedings of the International Conference "Manuscripts and Texts, Languages and Contexts: the Transmission of Knowledge in the Horn of Africa"*,

Wiesbaden, O. Harrassowitz, coll. « Supplement to Aethiopica. International Journal of Ethiopian and Eritrean Studies » 4, 281–295.

Guesdon, Marie-Geneviève. 2005. « Manuscrits de provenance yéménite donnés à la Bibliothèque nationale par Pierre Bardey en 1930 », *Chroniques yéménites* 13, 59–72.

Guesdon, Marie-Geneviève. Janvier 2013. « La Collection Colin à la Bibliothèque nationale de France », *Chroniques du manuscrit au Yémen* 15.

Humbert, Geneviève. 1997. « À la recherche de manuscrits dans les bibliothèques privées du Yémen : première mission à Zabid », *Chroniques yéménites* 96–97, 32–35.

Humbert, Geneviève. Novembre 2002. « Le manuscrit arabe et ses papiers », dans : Geneviève Humbert, *La tradition manuscrite en écriture arabe, Revue des mondes musulmans et de la Méditerranée* 99–100, 55–77 [mis en ligne le 12 mai 2009. URL : http://remmm.revues.org/1174].

Hunter, Dard. 1978. *Papermaking. The History and Technique of an Ancient Craft*, New York, Dover [repub. de la seconde éd. de 1947, revue et augmentée].

Ibrāhīm, Amīr Muḥammad Ṣādiq. 2010. *Al-ḫuṭūṭ wa-al-ʿalāmāt al-māʾiyya fī al-maḫṭūṭāt al-ʿarabiyya*, Riyad, Maktabat al-Malik ʿAbd al-ʿAzīz al-ʿāmma.

ʿĪsāwī, ʿIṣām Aḥmad. [2002]. *Al-waraq fī Miṣr fī al-qarn al-tāsiʿ ʿašar wa-ahamiyyatuhu fī al-dirāsāt al-diblūmāṯiyya wa-al-bibliūġrāfiyya*, Alexandrie, Dār al-ṯaqāfa al-ʿilmiyya.

Jones, Russell. 2010. “Indonesia and Malay Manuscripts and links to the Ottomans”, *Islam, Trade and Politics across the Indian Ocean*, British Academy. http://www.ottomansoutheastasia.org/ottoman-watermarks-in-malay-manuscripts.php.

Kâğıtçı, Mehmet Ali. 1976. *Historique de l'industrie papetière en Turquie*, Istanbul, Grafik Sanatlar Matbaası.

Al-Malīh, Muḥammad Saʿīd, & ʿAysawī, Aḥmad Muḥammad. [1978]. *Fihris maḫṭūṭāt al-Maktaba al-ġarbiyya bi-al-Ǧāmiʿ al-kabīr bi-Ṣanʿāʾ*, Alexandrie.

Mattozzi, Ivo. 1994. « I Galvani, fabbricanti di carta (1744–1855). Un modello di formazione dell'imprenditorialità? », dans : Gilberto Ganzer (a cura di), *Andrea Galvani, 1797–1855 : cultura e industria nell'Ottocento a Pordenone*, Pordenone, Studio Tesi, p. 17–41.

Messick, Brinkley. 1992. *The Calligraphic State. Textual Domination and History in a Muslim Society*, Berkeley/Los Angeles/Londres, Univeristy of California Press.

Nikolaev, Veslovod. 1954. *Watermarks of the Mediaeval Ottoman Documents in Bulgarian Libraries, Watermarks of the Ottoman Empire*, Sofia, Bulgarian Academy of Sciences, vol. 1 [titre original en bulgare].

O'Fahey, Rex Seán. 2003. *Arabic Literature of Africa*. Vol. 3, fasc. A : *The Writings of the Muslim Peoples of Northeastern Africa*, Leiden/Boston, E. J. Brill, « Handbook of Oriental Studies » section 1, The Near and Middle East, 13.

Phillips Paper Trade Directory of the World. 1906/1907–1946/1947. London, S. C. Phillips & Co.

Regourd, Anne. 2006. « Les routes commerciales entre Zabîd et l'Europe : les papiers filigranés de fonds manuscrits de Zabîd (Yémen, fin 18e–milieu 20e s.) », dans : Renato Traini, *Convegno Storia e cultura dello Yemen in età islamica con particolare riferimento al periodo rasûlide (Roma, 30–31 octobre 2003)*, Rome, Bardi, 173–198.

Regourd, Anne. (dir.). 2006. *Catalogue cumulé des bibliothèques de manuscrits de Zabid. I. Bibliothèque 'Abd al-Rahman al-Hadhrami*, fasc. 1, Sanaa, Centre français d'Archéologie et de Sciences sociales, Fonds social de Développement.

Regourd, Anne., avec la collaboration d'Hélène C. David et de 'Abd al-Rahman al-Ahmar. 2008. *Catalogue cumulé des bibliothèques de manuscrits de Zabid. I. Bibliothèque 'Abd al-Rahman al-Hadhrami*, fasc. 1, *Les papiers filigranés*, Sanaa, CEFAS, FSD.

Regourd, Anne. (dir.). 2009. *Catalogue cumulé des bibliothèques de manuscrits de Zabid. I. Bibliothèque 'Abd al-Rahman al-Hadhrami*, fasc. 2, Sanaa, CEFAS, FSD.

Regourd, Anne. Février 2009. « L'inventaire du fonds manuscrit de la fondation Hâ'il Sa'îd (Taez) », *Chroniques du manuscrit au Yémen* 7 (avec inventaire).

Regourd, Anne. 2011. « Images de djinns et exorcisme dans le *Mandal al-sulaymani* » (avec éd. de texte et trad.), dans : Anna Caiozzo, Jean-Patrice Boudet, Nicolas Weill-Parot (éd.), *Autour de Picatrix : Images et magie*, Proceedings of the International Congress, Institut national d'histoire de l'art, Paris, 2007, Paris, Champion.

Regourd, Anne. 2014. « Introduction to the Codicology: Papers, Ruling, and Bindings », dans : Alessandro Gori & coll., voir à cette entrée, xlix–xcv.

Regourd, Anne, (dir.). À paraître. *Catalogue cumulé des bibliothèques de manuscrits de Zabid. I. Bibliothèque 'Abd al-Rahman al-Hadhrami*, fasc. 3.

Regourd, Anne. À paraître. « Paper Trade on the Red Sea (XIXe c.–first half of the XXe c.): 'local papers' against Italian's. The case of an 'Ottoman' watermarked paper used in Yemen and in Ethiopia », *Festschrift in Honour of Professor Iraj Afshar*, Londres, Fondation al-Furqān.

Regourd, Anne. À paraître. *Handlist of the Manuscripts in the 'Abd Allah Sharif Museum, Harar*, Eugene (OR), Pickwick Publication, « Ethiopian Manuscrits, Texts, and Studies Series », fasc. 1.

Rossi, Ettore. 1938. « La Stampa nel Yemen », *Oriente Moderno* 18, 568–580.

Serjeant, Robert Bertram. 1967. *The Sayyids of Hadhramaut*, Londres, School of Oriental and African Studies.

Serjeant, Robert Bertram. 1962. « History and Historiography of Hadhramaut », *Bulletin of School of Oriental and African Studies* XXV, 239–471.

Serjeant, Robert Bertram & Ronald Lewcock. 1983. *Ṣan'ā'. An Arabian Islamic City*, Londres, World of Islam Festival Trust [1re imp.].

Shuqayr, Na'um. 1967. *Ǧuġrafiyya wa-tārīḫ al-Sūdān*, Beyrouth.

Tang Abdullah, Abdul Rahman. 2009. « Arab Hadhramis in Malaysia: their origins and assimilation in Malay society », dans : Ahmed Ibrahim Abushouk & Hassan Ahmed Ibrahim, *Hadhrami Diaspora in Southeast Asia. Identity Maintenance or Assimilation?*, Leiden/Boston, E. J. Brill, 45–56.

Tekin, Şinasi. 1993. *Eski türklerde yazı, kâğıt, kitap ve kâğıt damgaları*, Istanbul.

Velkov, Asparoukh, & Andreev, Stephane. 1983. *Filigranes dans les documents ottomans.* Vol. 1 : *Trois croissants*, Sofia, « Bibliothèque Cyrille et Méthode », Série Orientale [en bulgare et français].

Velkov, Asparoukh. 2005. *Les filigranes dans les documents ottomans. Divers types d'images*, Sofia, éd. « Textes – A. Trayanov ».

Voorhoeve, Petrus. 1980. *Handlist of Arabic manuscripts in the library of the University of Leiden and other collections in the Netherlands*, La Haye/Boston, Leiden University Press [2e éd.].

Walz, Terence. 1985. « The Paper Trade of Egypt and the Sudan in the Eighteenth and Nineteenth Centuries », dans : Martin William Daly (éd.), *Modernization in the Sudan: Essays in Honor of Richard Hill*, New York, Lillian Barber Press, Inc., 29–48. Version révisée : Walz, Terence. 2011. « The paper trade of Egypt and the Sudan in the eighteenth and nineteenth centuries and its re-export to the Bilād as-Sūdān », dans : Graziano Krätli & Ghislaine Lydon, *The Trans-Saharan Book Trade: Manuscript Culture, Arabic Literacy and Intellectual History in Muslim Africa*, Leyde, E. J. Brill, 73–107.

Annexe

Ms. UbL Or. 6634, daté au colophon du 24 ḏū al-ḥiǧǧa 1346/13.06.1928. Papiers A.1. et A.2., formés en quaternions (voir I. A.1., A.2.1., A.2.2. etc.). Le type A.1. est dominant, il y a peut-être un cahier mixte. Les regroupements de cahiers suivant reprennent des différences sensibles, externes ou internes (par transparence) entre les papiers de type A.1. :

cahiers 1–7, Abū Šubbak A.1. ; cahiers 8–14, Abū Šubbak A.2. ; cahier 15 peut-être mélangé (2 bifeuillets de A.2. et 2 de A.1.) ; cahiers 16–20, Abū Šubbak A.1. ; cahiers 21–28, Abū Šubbak A.1.

L'éventail d'images restitue partiellement ces différences sensibles que les mots rendent difficilement ou de manière imprécise.

ILLUSTRATION ANNEXE 1 *Bloc des cahiers. Ms. UbL Or. 6634, daté 1346/1928*

ILLUSTRATION ANNEXE 2 *Type A.1. Contremarque. Ms. UbL Or. 6634, daté 1346/1928*

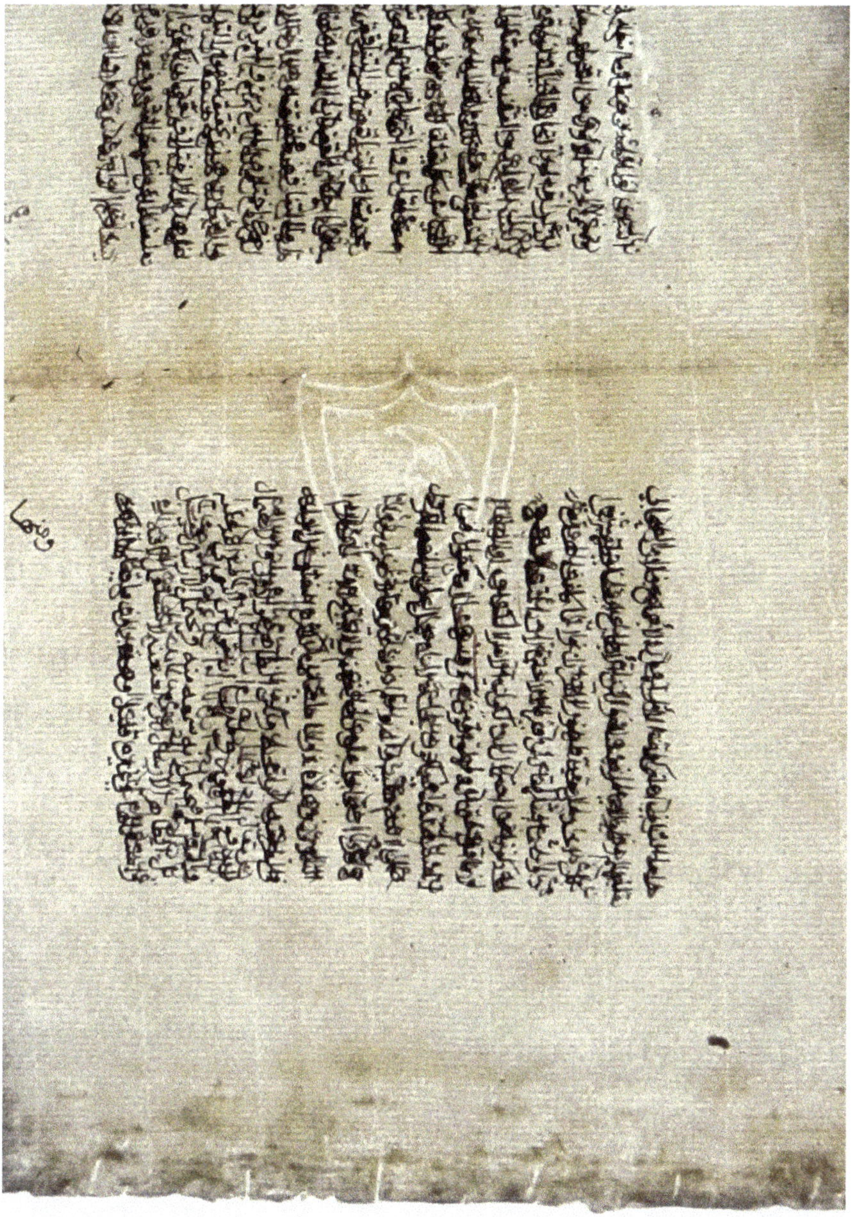

ILLUSTRATION ANNEXE 3 *Type A.1. Écu. Ms. UbL Or. 6634, daté 1346/1928*

ILLUSTRATION ANNEXE 4 *Type A.2. Écu. Ms. UbL Or. 6634, daté 1346/1928*

ILLUSTRATION ANNEXE 5 *Type A.2. Écu. Ms. UbL Or. 6634, daté 1346/1928*

ILLUSTRATION ANNEXE 6 *Type A.2. Contremarque. Ms. UbL Or. 6634, daté 1346/1928*

ILLUSTRATION ANNEXE 7 *Bord. Ms. UbL Or. 6634, daté 1346/1928*

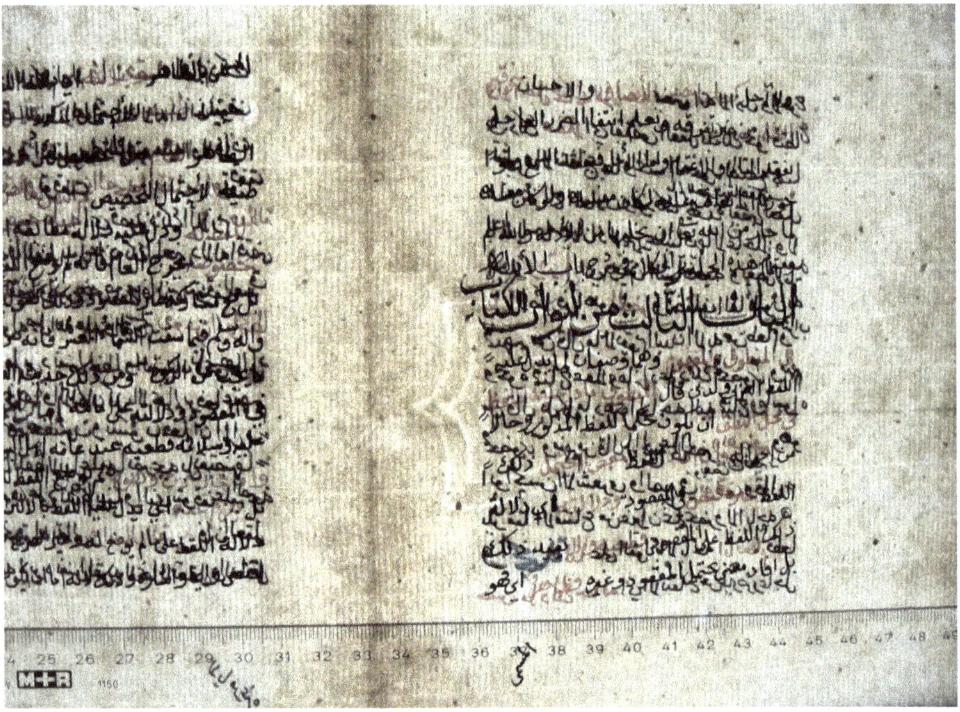

ILLUSTRATION ANNEXE 8 *Type A.1. Écu. Ms. UbL Or. 6634, daté 1346/1928*
© AIMABLE AUTORISATION UBL

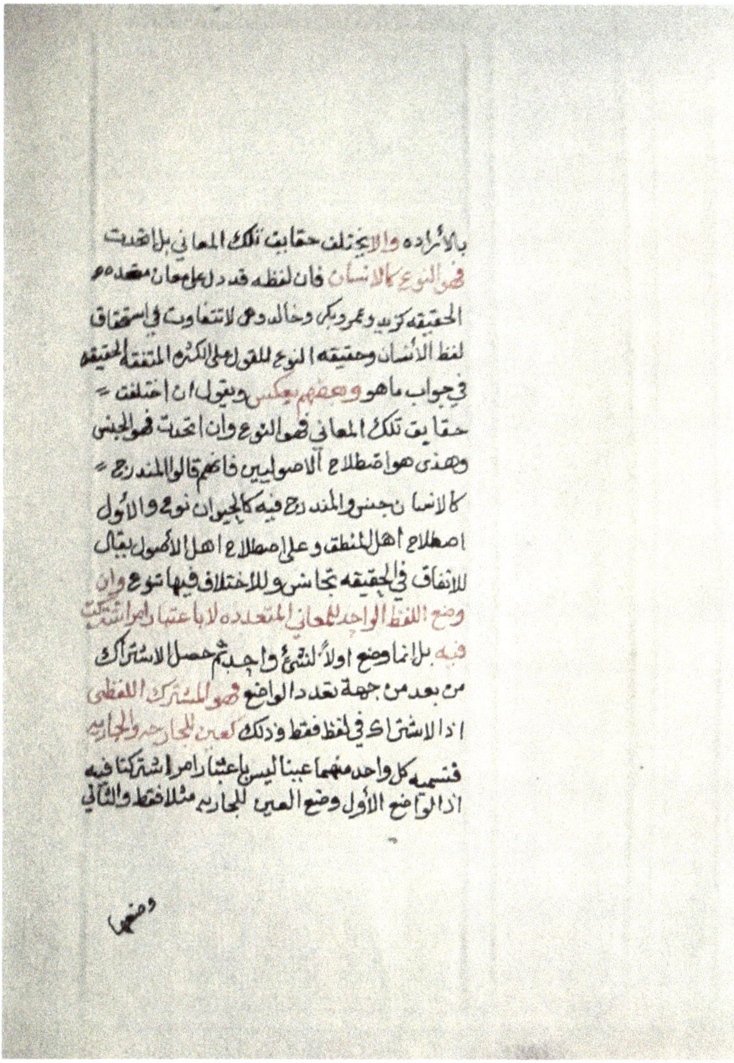

بالأراده والايختلف حقايق تلك المعاني بل اتحدت
فهو النوع كالانسان فان لفظه قد دل على معان متعدده
الحقيقه كزيد وعمرو وبكر وخالد وهي لاتتفاوت في استحقاق
لفظ الانسان وحقيقه النوع للقول على الكثرة المتفقة الحقيقه
في جواب ماهو وبعضهم يعكس ويقول ان اختلفت
حقايق تلك المعاني فهو النوع وان اتحدت فهو الجنس
وهذى هو اصطلاح الاصوليين فانهم قالوا المندرج
كالانسان جنس والمندرج فيه كالحيوان نوع والاول
اصطلاح اهل المنطق وعلى اصطلاح اهل الاصول يقال
للاتفاق في الحقيقه تجانس وللاختلاف فيها تنوع وان
وضع اللفظ الواحد للمعاني المتعدده لا باعتبار امر اشتركت
فيه بل انما وضع اولاً لشيء واحد ثم حصل الاشتراك
من بعد من جهة تعدد الواضع فهو المشترك اللفظي
اذ الاشتراك في لفظ فقط وذلك كعين للجارحة والجاريه
فتسميه كل واحد منهما عينا ليس باعتبار امر اشتركتا فيه
اذ الواضع الاول وضع العين للجاريه مثلا فقط والثاني

وضعها

ILLUSTRATION ANNEXE 9 *Apparence externe. Ms. UbL Or. 6634, daté 1346/1928*

ILLUSTRATION ANNEXE 10 *Type A.1. Contremarque. Ms. UbL Or. 6634, daté 1346/1928*

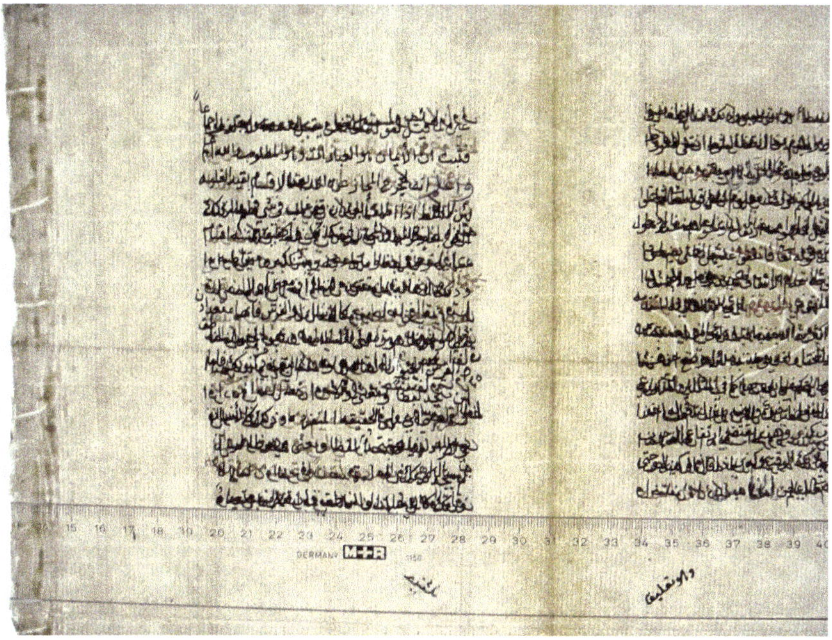

ILLUSTRATION ANNEXE 11 *Type A.1. Écu. Ms. UbL Or. 6634, daté 1346/1928*

ILLUSTRATION ANNEXE 12 *Bord. Ms. UbL Or. 6634, daté 1346/1928*

ILLUSTRATION ANNEXE 13 *Type A.1. Écu. Ms. UbL Or. 6634, daté 1346/1928*
© AIMABLE AUTORISATION UBL

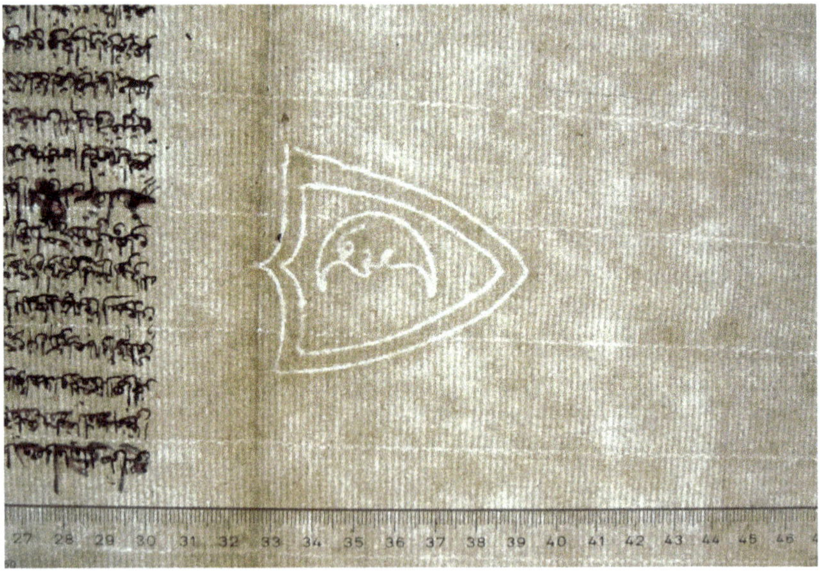

ILLUSTRATION ANNEXE 14 *Type A.1. Écu. Ms. UbL Or. 6634, daté 1346/1928*
© AIMABLE AUTORISATION UBL

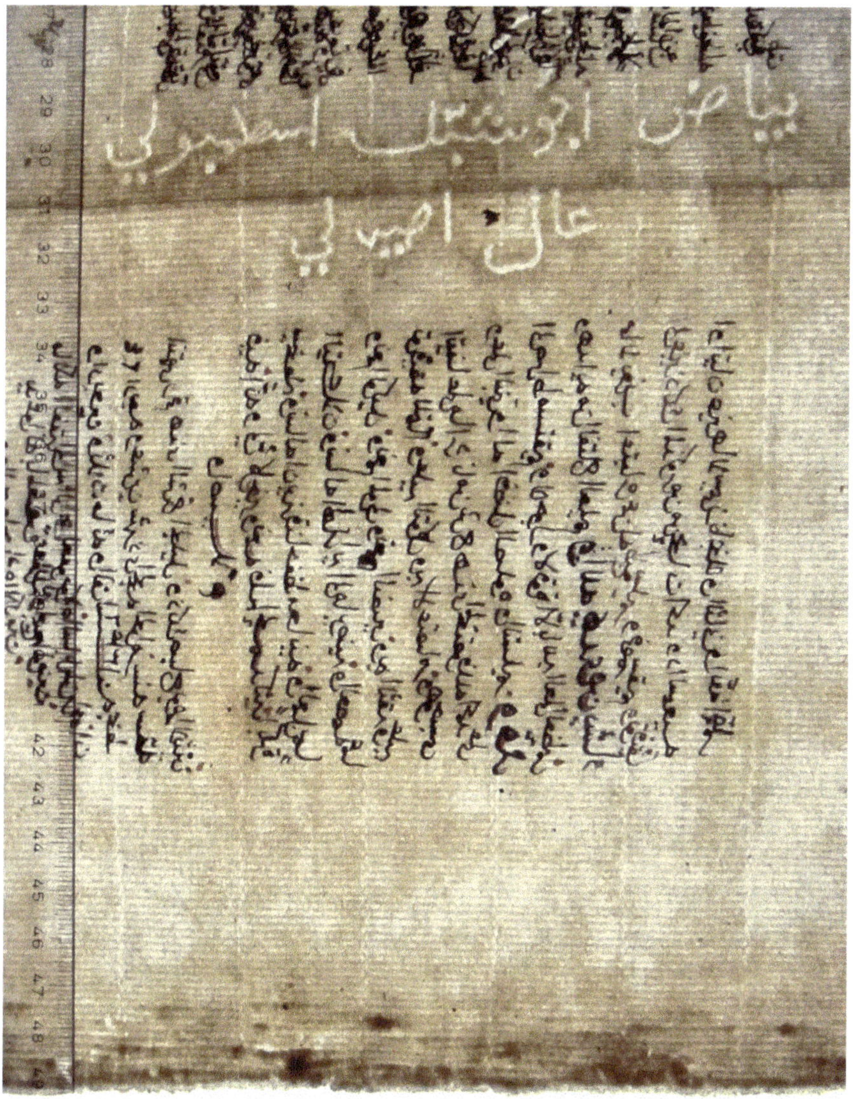

ILLUSTRATION ANNEXE 15 *Type A.1. Contremarque. Ms. UbL Or. 6634, daté 1346/1928*
© AIMABLE AUTORISATION UBL

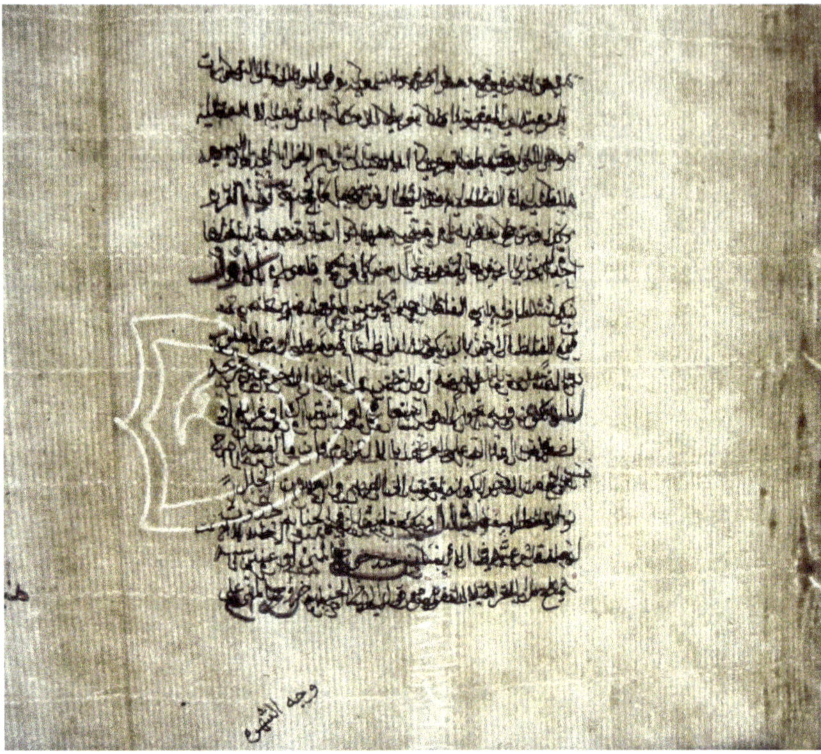

ILLUSTRATION ANNEXE 16 *Type A.1. Écu. Ms. UbL Or. 6634, daté 1346/1928*
© AIMABLE AUTORISATION UBL

ILLUSTRATION ANNEXE 17 *Bord. Ms. UbL Or. 6634, daté 1346/1928*
© AIMABLE AUTORISATION UBL

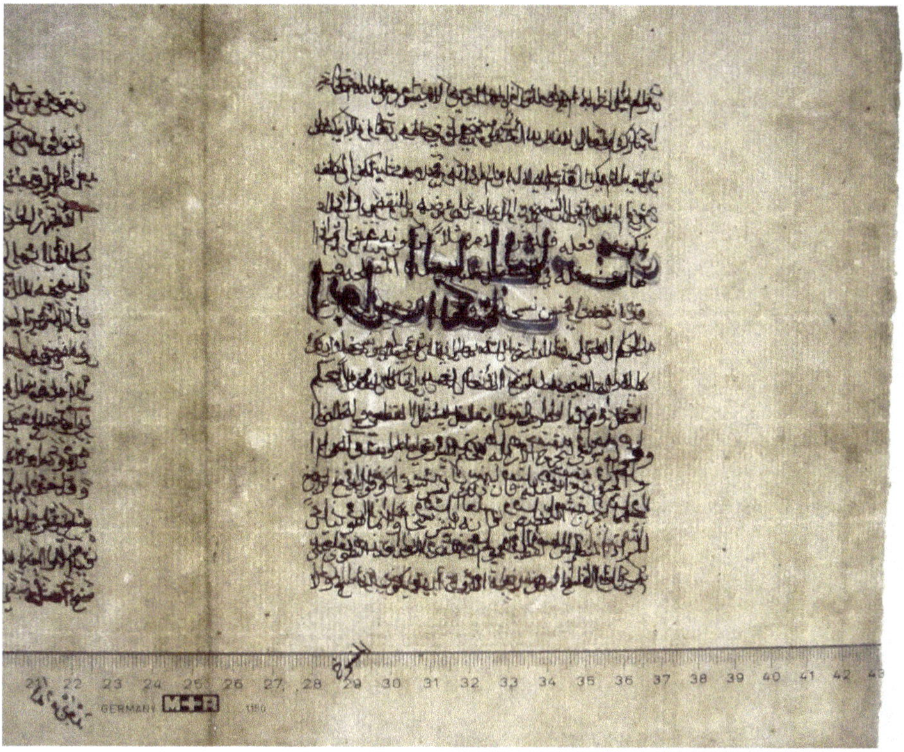

ILLUSTRATION ANNEXE 18 *Type A.1. Écu. Ms. UbL Or. 6634, daté 1346/1928*

ILLUSTRATION ANNEXE 19 *Type A.1. Contremarque. Ms. UbL Or. 6634, daté 1346/1928*

ILLUSTRATION ANNEXE 20 *Bord. Ms. UbL Or. 6634, daté 1346/1928*

CHAPTER 5

Papiers « indiens » de manuscrits éthiopiens (fin xixe–début xxe siècle)[1]

Anne Regourd

Abstract

We showed in a previous study that a paper "Bombay", which had come to Yemen, probably through Aden, was manufactured in England and then distributed by an Indian merchant. The hypothesis was put forward that the network of these Indian distributors extended to Ethiopia (Harar), albeit indirectly. A series of papers discovered in the spring of 2015 allows us to confirm the presence in the city of Harar of papers manufactured in the West for Indian merchants, and to attempt to reconstruct their trade route.

1 Introduction

Dans le registre IES 963 de l'Institut d'études éthiopiennes (IES) de l'Université d'Addis Abeba, provenant de Harar, se trouve un document volant en arabe écrit sur un bifeuillet portant une contremarque en caractères latins et arabes, mentionnant Bombay.

Loin d'être un cas isolé, d'autres documents volant en arabe, avec, parfois, quelques mentions en amharique, conservés dans ce registre et dans un autre venant aussi de Harar, ont un papier à contremarques latines indiquant

1 Ces papiers ont été principalement réunis lors d'une mission du Projet European Research Council (ERC) « Islam in the Horn of Africa. A Comparative Literary Approach », Advanced Grant no. 322849, Université de Copenhague, que j'ai effectuée à Addis Abeba en avril 2015. Je tiens à remercier le Directeur de l'Institute for Ethiopian Studies (IES), à Addis Abeba, Ahmad Hassan, ainsi que les bibliothécaires, tout particulièrement Woolet Selassie, pour leur gentillesse et leur patience. Sara Fani et Michele Petrone, ERC « Islam in the Horn of Africa », ont aimablement collecté les papiers de la section IV, « Papiers additionnels de codex du pays Oromo » à mon attention, lors de leur mission du printemps 2016 : qu'ils trouvent ici l'expression de ma reconnaissance pour leur geste amical. Sara Fani est l'auteur des photographies dont elle a bien voulu autoriser la reproduction.

 | DOI 10.1163/9789004360877_007

des connexions avec l'Inde. En ont été sélectionnés ici les exemples les plus complets.

Datés dans le calendrier éthiopien, avec quelquefois la correspondance en calendrier de l'hégire, ils font écho à un papier aux marques similaires trouvé dans un codex de Zabid (Yémen)[2]. L'ensemble représente un corpus intéressant l'étude des routes du papier.

2 Données textuelles et description du papier

Le document a été trouvé entre les pages 97 et 98 du registre IES 963.

Texte
2 dates en 1911/1918–1919 ; à l'intérieur, sur l'un des feuillets, il est question entre autre de maïs/mil « ʿan bayān al-ḥabb al-durra » ; à l'extérieur, se trouvent 7 l. et 1 mot en amharique, tête-bêche[3]. Des colonnes ont été tracées au crayon à mine.

Papier machine. Il est fin, d'apparence jaune clair, il se déchire aisément. Il est pré-ligné (lignes bleues) sans marque de marge.

Dimensions maximales du bifeuillet : 33,7 × 42 cm
Distance entre 2 fils de chaînette : 2,5 cm
20 vergeurs : 2,7 cm

à gauche du bifeuillet, de haut en bas :

- Contremarque en anglais/caractères latins

hauteur (reconstituée) : 13,5 cm
largeur maximale : 14,3 cm

2 A. Regourd, « Papiers filigranés de manuscrits de Zabīd, premier tiers du XVIIIe-milieu du XXe siècle : papiers importés et 'locaux' », 2015.

3 Les indications textuelles servent deux buts : permettre au chercheur d'identifier le document, sur feuille volante, et d'identifier le (ou les) type(s) de document pour le(s)quel(s) ce papier a été utilisé.

MAHOMEDALLY[BHOY ?] WALLY IARVARI	(lettres doubles, semi-circulaire)
N° 68	(lettres doubles, centré, à l'intérieur de l'inscription semi-circulaire)
SUTAR SHAWL	(lettres doubles, centré, à l'intérieur de l'inscription semi-circulaire)
BOMBAY	(lettres doubles, centré, à l'intérieur de l'inscription semi-circulaire)
BRITISH MAKE	(rectiligne, hors champ inscription semi-circulaire)
BIN (?) HUSI (?)	(lettres doubles)

Au-dessous, à une distance de 5,5 cm :

- Filigrane

hauteur maximale : 6 cm
largeur maximale : 5,2–3 cm

Étoile à 8 branches

Au-dessous, en caractères arabes :

- Contremarque en caractères arabes

Hauteur maximale : 6,3 cm (l'*alif* de « stār »parvient à hauteur de l'étoile)
largeur maximale : 14,3 cm

ســــــتار في سند

soit : « stār fī Sind », l'étoile du Sind.
Le mot « stār » domine par ses dimensions. On note le *sīn* non-indenté de *stār*, le *fī* dont le *yā'* revient sous le *fā'*, qui a son point diacritique placé à gauche de la boucle.

à droite du bifeuillet, de haut en bas

- Contremarque à caractères arabes

mêmes texte et caractéristiques graphiques

ســـــتار في سند

Au-dessous, à une distance de 3,3 cm :

- Contremarque en caractères latins

hauteur lettres : 1,7 cm
longueur maximale : 9,8 cm

KUTUB TA (lettres doubles)

N.B. : Il est probable que le même texte apparaît sous l'autre occurrence, qui est placé trop près du bord pour qu'il ait une chance d'apparaître.

Au-dessous, à une distance de 5,5 cm :

- Filigrane

Étoile : hauteur × largeur maximales : 5,3 cm
Anges : hauteur maximale, jusqu'au bord inférieur du bifeuillet (filigrane incomplet) : > 14,5 cm ; les 2 anges occupent à peu près la largeur d'1 feuillet

Étoile à 8 branches, tenue par deux anges.

Seule la contremarque en caractères arabes apparaît à droite et à gauche du bifeuillet, mais décalée, en alternance avec la contremarque en caractères latins et le filigrane aux anges [Ill. 5.1].

On compte une autre occurrence de ce papier parmi les documents volants entre les pages du registre, mais le texte ne comporte aucune date et les dimensions du document sont nettement plus petites (11,5 × 10,3 cm).

Loin d'être un cas isolé, d'autres documents sur papiers dont les marques se réfèrent à l'Inde, surtout à Bombay, ont été trouvés.

3 Autres exemples

Onze autres exemples ont été réunis en avril 2015.

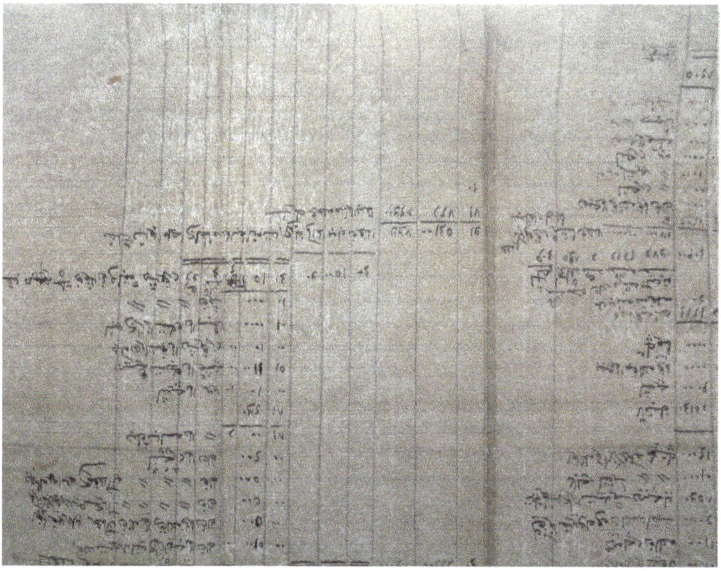

ILLUSTRATION 5.1A *Papier au nom de* MAHOMEDALLY*[*BHOY *?]* WALLY IARVARI*.*
Partie supérieur gauche du bifeuillet

PHOTOGRAPHIE ANNE REGOURD

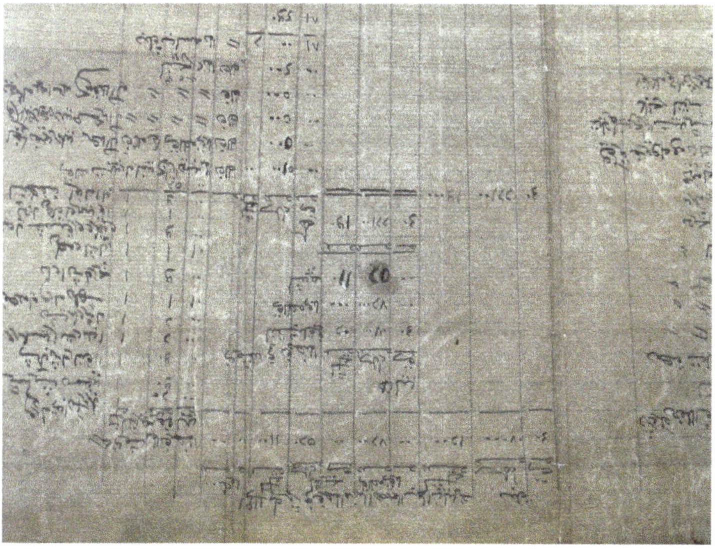

ILLUSTRATION 5.1B *Papier au nom de* MAHOMEDALLY*[*BHOY *?]* WALLY IARVARI*.*
Partie inférieure gauche du bifeuillet

PHOTOGRAPHIE ANNE REGOURD

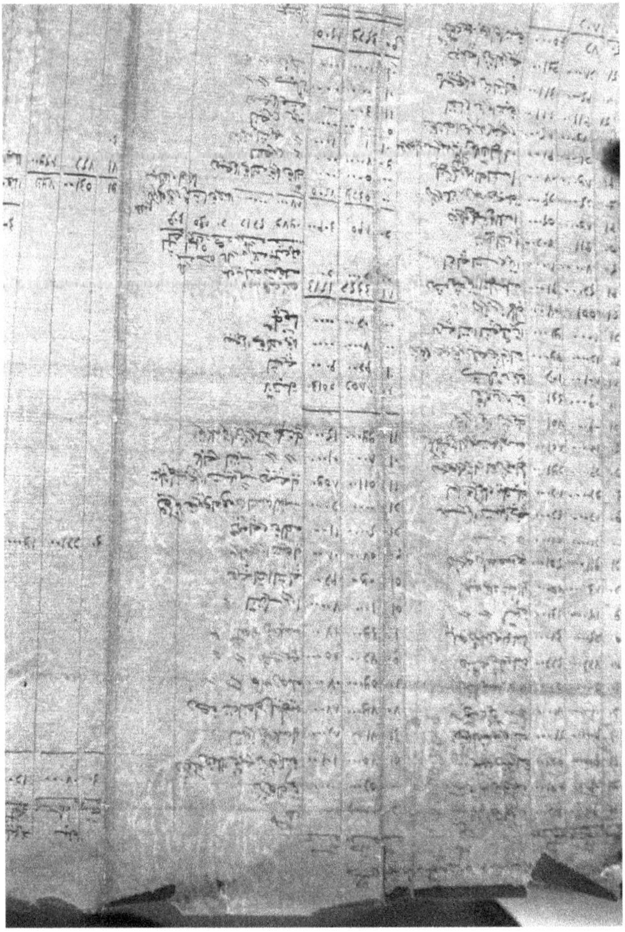

ILLUSTRATION 5.1C *Papier au nom de* MAHOMEDALLY[BHOY ?] WALLY IARVARI. *Partie droite du bifeuillet* © AIMABLE AUTORISATION DE L'IES, ADDIS ABEBA. PHOTOGRAPHIE ANNE REGOURD

3.1 *Registre* IES *962*

Il s'agit d'un registre en arabe de la cour (*šarīʿa*) du Harar, établi durant la période d'occupation par l'Égypte, couvrant de *ṣafar* 1299 à *ǧumādā al-awwal* 1302/décembre 1881 à février 1885. Il comporte 61 feuillets, à raison de 35–44 lignes par page. Les dimensions du registre sont de 45,8 × 31,5 × 1,5 cm[4].

4 A. Gori, *A Handlist of the Manuscripts in the Institute for Ethiopian Studies. The Arabic Material of the Ethiopian Islamic Tradition*, 2014, IES 962, دفتر الادارة المصرية في هرر.

3.1.1 *Bombay, Ahmedabad* : PATHAK SANGAVI

Dans le registre IES 962, notons un document lié à notre recherche, glissé entre le contreplat et la première page (non-paginée). Il s'agit d'un bifeuillet, qui porte également une pliure horizontale volontaire à mi-hauteur.

Texte
S. d. ; chaque feuillet interne porte en colonne une liste de noms, précédés par des chiffres ; la colonne de droite est introduite par : « ḍamānat Ǧarrād Adam Ibrāhīm », celle de gauche par : « ḍamānat Yūsuf Saʿīd » ; au-dessus des chiffres : « … / ʿadad ». Des calculs émaillent l'espace demeuré libre.
Verso vierge.

Papier machine. Il est blanc d'apparence, jauni (taches) ; bords supérieurs et inférieurs ont été coupés avec un instrument tranchant. Il a été pré-ligné (lignes bleues).

Dimensions maximales du bifeuillet : 33,8 × 21 cm
Distance entre 2 fils de chaînette : 2,5 cm
20 vergeurs : 2,8 cm

- Marque de fabrique, caractères latins

BRITISH MAKE	(lettres doubles, rectiligne)
FOR	(lettres doubles, rectiligne)
PATHAK SANGAVI & [CO ?]	(semi-circulaire convexe)
BOMBAY & AHMED[ABAD ?	(inscrit dans le demi-cercle du texte supérieur, rectiligne)

Une partie du texte manque car le bifeuillet est déchiré ; par symétrie, on peut calculer qu'il manque 2,5 à 3 cm de texte.

3.1.2 *Non localisé* : MAHOMEDALLY …

Il s'agit d'un bifeuillet prélevé d'un cahier, car plié au centre, il porte des points des coutures.

Texte
Date : 10 rabīʿ awwal 1324/4 mai 1906.
Verso vierge.

Papier machine ; jaune-brun ; pré-ligné.

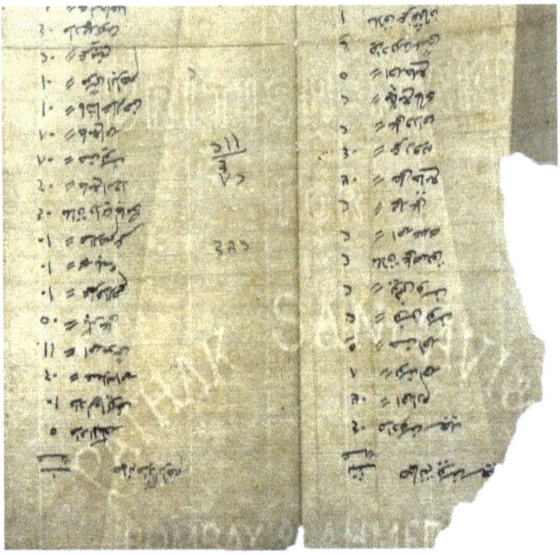

ILLUSTRATION 5.2
Papier au nom de
PATHAK SANGAVI & [*CO ?*]
© AIMABLE AUTORISATION
DE L'IES, ADDIS ABEBA.
PHOTOGRAPHIE ANNE REGOURD

Dimensions maximales du document : 33 × 42 cm
Distance entre 2 fils de chaînette : 2,5 cm
20 vergeurs : 2,8 cm

- Contremarque (à gauche)

MAHOMEDALLY ARD.........OSEM (?) (semi-circulaire)
MADE IN ENGLAND.

- Filigrane (à droite)

navire à voiles voguant sur des flots, au sommet du bateau, un coq, petits nuages dans le ciel ;
à gauche : VACHI ; à droite YAHOT[5].

la ligne du filigrane paraît discontinue.

3.2 *Registre* IES *963*

Il s'agit d'un autre registre administratif, « Don de Harar ». Le mot *dīwān* est imprimé sur la couverture. Des timbres secs, présents dans le coin supérieur droit, authentifient ce qui est enregistré. Le registre couvre les années 1890 à

5 Des familles s'appellent Vachi et Yahot.

1892 (calendrier éthiopien), 1315–1316 à 1317–1318 de l'Hégire et 1897 à 1899 de notre ère.

En dehors du papier « Stār fī Sind » (II.), 8 autres documents sur papiers dont les marques se réfèrent à l'Inde ont été identifiés. Tous sont des feuillets volants, conservés entre les pages du registre. Ces papiers peuvent être divisés en 3 groupes.

3.2.1 Papiers de fabrication étrangère

Sont identifiables des marchands de Bombay. Ils commandent leurs papiers aussi bien en Angleterre qu'en Belgique.

3.2.1.1 *Bombay : papiers* SHAIK AHMED SHAIK DAWOOD

Différentes contremarques au nom de ce marchand de Bombay ont été notées. Deux informations intéressantes y figurent : les papiers sont commandés non seulement en Angleterre, mais aussi en Belgique ; l'une des contremarques donne la date de 1899.

3.2.1.1.1 Fait en Angleterre

Le document se trouvait entre les pages 75 et 76 du registre IES 963.

Texte

S. d. ; 3 colonnes de noms introduites par : « ʿan bayān ʿadad al-ʿasākir al-musāfirīn ilā ǧihat Ǧubūt bi-rifqa / anfār al-ʿasākir / ʿadad »[6] ; de gauche à droite, en haut de chaque colonne, figurent : « qabīlat Nūlī », « qabīlat Āl Zaḥṭan », « qabīlat Ǧār S… tabaʿ Ḥasan Bašīr » ; au tiers des colonnes 2 et 3 : « Diyāb Afandī, ʿUmar (…) bi-yadihi » et « Ṭayḥī ». Certains noms ont été biffés, la plupart corrigés. Une marque en forme de croix est placée après les noms, certainement pour marquer les présents. Les personnes sont décomptées.

Verso vierge.

Si le texte ne livre aucune date, la contremarque du papier indique 1899, certainement de notre ère. Le feuillet est plié en accordéon dans le sens vertical.

Papier machine.

Dimensions maximales du document : 33,6 × 31,6 cm

- Contremarque

6 Les barres obliques indiquent les changements de ligne.

MADE IN ENGLAND	(rectiligne)
FOR	(rectiligne)
SHAIK AHMED SHAIK DAWOOD	(semi-circulaire convexe)
OF	(centré, inscrit dans le demi-cercle du texte supérieur, rectiligne)
BOMBAY	(centré, inscrit dans le demi-cercle du texte supérieur, rectiligne)
1899	(centré, inscrit dans le demi-cercle du texte supérieur, rectiligne)

- Filigrane

Britannia

Assise sur un trône avec une croix et à ses pieds les flots, elle tient obliquement un trident dont le bout du manche repose sur le « sol », elle porte un casque à crinière, dans un médaillon dessiné par un double trait et surmonté par une couronne avec une croix ; à l'intérieur de la couronne, autre croix, dans l'axe de la première.

- Disposition du nom du marchand/revendeur/transporteur et du filigrane sur le bifeuillet

16,5 cm bord supérieur *id.*
^
| 2 cm – Nom revendeur – 9,5 cm – Filigrane [
^
4,5 cm bord inférieur *id.*

3.2.1.1.2 Fait en Angleterre, variante (contremarque)

Texte
S. d. ; dans une direction : « riyāl ǧimāl ʿidda ṣanādīq tammat », dessous « ʿadad » pour chaque colonne et liste de noms ; dans une autre direction : « ǧimāl ṣanādīq », dessous « ʿadad » pour chaque colonne et liste de noms.
Verso vierge.

Papier machine. Apparence blanche.
Dimensions maximales du document : 21 × 16,9 cm
Distance entre 2 fils de chaînette : 2,5 cm
20 vergeurs : 2,8 cm

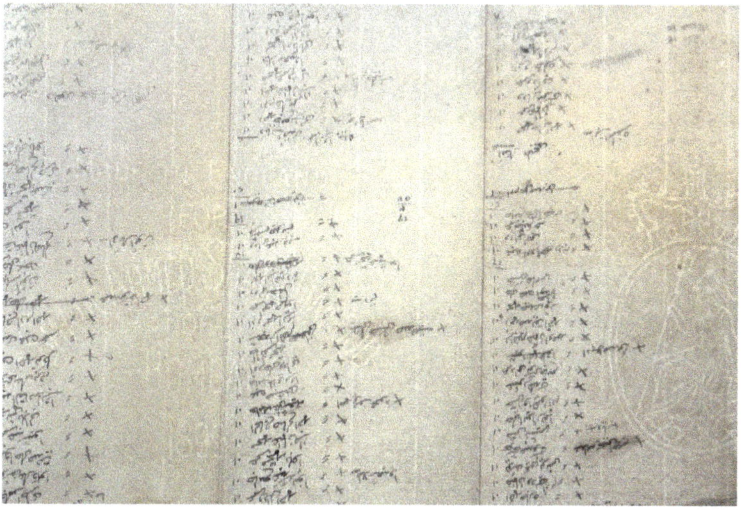

ILLUSTRATION 5.3A *Papier au nom de* SHAIK AHMED SHAIK DAWOOD
© AIMABLE AUTORISATION DE L'IES, ADDIS ABEBA. PHOTOGRAPHIE ANNE REGOURD

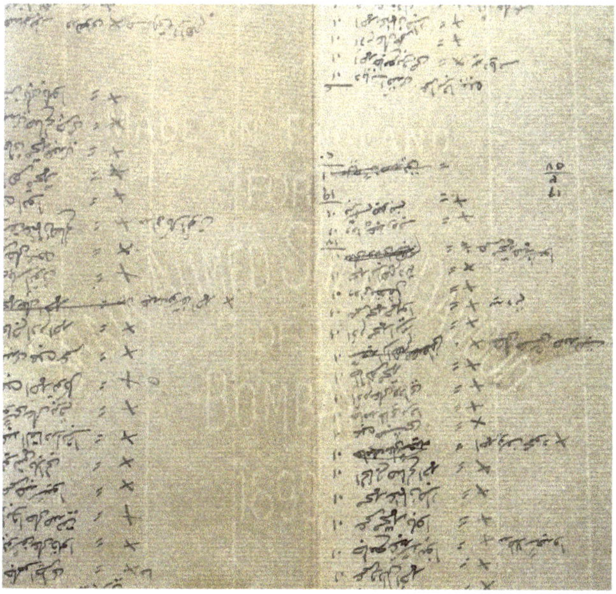

ILLUSTRATION 5.3B *Papier au nom de* SHAIK AHMED SHAIK DAWOOD. *Contremarque*
© AIMABLE AUTORISATION DE L'IES, ADDIS ABEBA. PHOTOGRAPHIE ANNE REGOURD

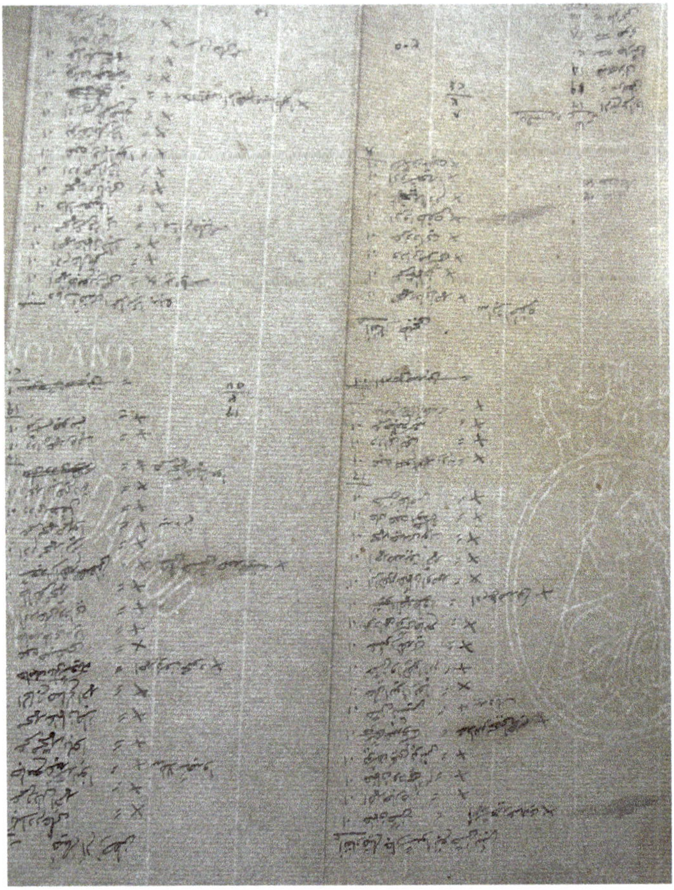

ILLUSTRATION 5.3C *Papier au nom de* SHAIK AHMED SHAIK DAWOOD. *Filigrane*
© AIMABLE AUTORISATION DE L'IES, ADDIS ABEBA. PHOTOGRAPHIE ANNE REGOURD

- Contremarque

MADE IN ENGLAND	(semi-circulaire convexe)
FOR	(centré, rectiligne)
SHAIK AHMED SHAIK DAWOOD	(semi-circulaire concave)

La compagnie est enregistrée au Phillips' Paper Trade Directory of the World, pour l'année 1923, dans la section des marchands de papiers de Bombay, à l'entrée « Dawood (Shaik Ahmed Shaik) », 29, Sarang Street, comme importateurs et revendeurs (*dealers*) de toutes sortes de papiers, fournitures, etc.[7].

L'un des papiers à contremarque Shaikh Ahmad Shaikh Dawood de Bombay fabriqués en Angleterre a été utilisé pour la rédaction de codex en écriture *khojkī* : 3 venant de la communauté ismaélienne de l'Inde de l'Ouest (Kutch et Gudjerat), datés de 1890, sont présentés par Ali Sultaan Asani[8] ; 1, estimé des alentours de 1900, est conservé à Harvard[9].

3.2.1.1.3 Fait en Belgique

Document trouvé entre les pages 66 et 67 du registre IES 963.

Texte

Date : 22 šaʿabān 1310/11 mars 1893 ; début, « ʿan bayānat ʿaddad al-ʿafaš tuʿalliq ... al-musāfirīn (...) ».

Verso vierge.

Papier machine. Apparence blanche.

Dimensions maximales du document : 12 × 21 cm

Distance entre 2 fils de chaînette : 2,5 cm

20 vergeurs : 2,8 cm

- Contremarque

MADE IN BELGIUM	(lettres doubles, semi-circulaire convexe)
FOR	(lettres doubles, centré, rectiligne)
SHAIK AHMED SHAIK DAWOOD	(lettres doubles, semi-circulaire concave)

7 *Phillips Paper Trade Directory of the World* 1923, 526.

8 Il s'agit des mss. Ism K 6, Ism K 8 et Ism K 13, dont le papier a été décrit par A. S. Asani, *Ecstasy and Enlightenment: The Ismaili Devotional Literature of South Asia*, 2002, 136, « Paper and Watermarks », chap. « Khojkī and its manuscript tradition ». Au sujet de l'écriture *khojkī*, voir Z. Bhalloo & A. Iqbal, « Les manuscrits du sud de la vallée de l'Indus en écriture khojkī sindhī : état des lieux et perspectives », à paraître, sections 2 et 3.2.

9 Ms. Indic 2534, Bibliothèque Houghton, Université de Harvard, numérisé dans le cadre du « Islamic Heritage Project », Harvard, sans titre, contenant divers textes dévotionnels chiites liés au martyr de l'Imam Husayn, en hindustani, écriture Khojki, s. d., estimé des alentours de 1900, 21 × 17 cm, cf. Islamic Heritage Project, http://ocp.hul.harvard.edu/dl/ihp/011919186.

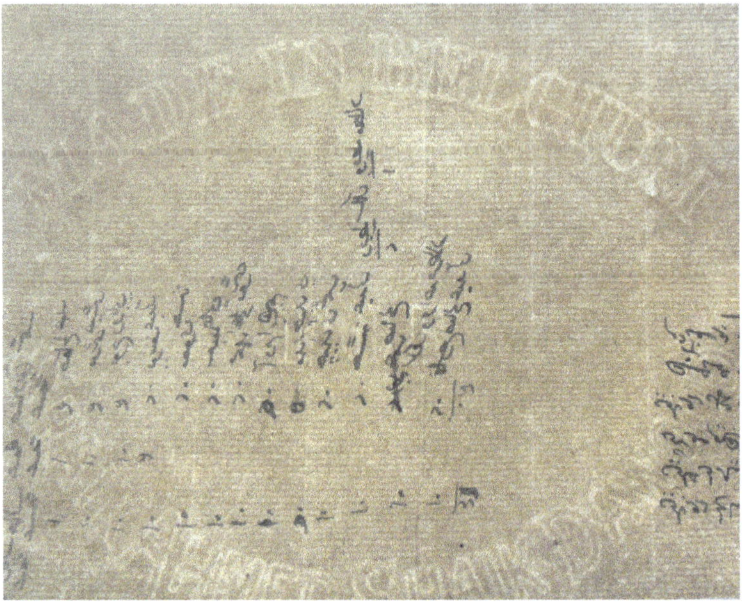

ILLUSTRATION 5.4 *Papier au nom de* SHAIK AHMED SHAIK DAWOOD
© AIMABLE AUTORISATION DE L'IES, ADDIS ABEBA.
PHOTOGRAPHIE ANNE REGOURD

3.2.1.2 *Bombay* : *NUZZURALLY HEBTOOLABHOY*

Le document décrit ici était placé entre les pages 31 et 32 du registre IES 963. La feuille a été pliée en 4 verticalement.

Texte

Date : 1891/1898–1899 ; début, partie externe du bifeuillet : « ʿasākir al-ḫāriǧīn ilā Ǧimma (…) » ; suivent 4 colonnes de noms, avec, en bas à gauche, d'autres calculs. Le texte se poursuit sur la partie interne, en 3 colonnes.

Papier

Papier machine. D'apparence jaune clair[10].
Dimensions maximales du document : 33,1 × 41,2 cm
Distance entre 2 fils de chaînette : 2,2 cm
20 vergeurs : 2,6 cm

10 Le bord inférieur d'un autre document correspond à celui d'une feuille originale. Le papier a légèrement bruni et se déchire facilement.

- Contremarque

MADE IN BELGIUM	(rectiligne)
FOR	(lettres doubles, centré, rectiligne)
NUZZURALLY HEBTOOLABHOY	(lettres doubles, semi-circulaire convexe)
BOMBAY	(lettres doubles, centré, inscrit dans le demi-cercle du texte supérieur, rectiligne)

Le texte de la contremarque a été recomposé, MADE IN BELGIUM apparaît en bas de la feuille. Même observation pour le filigrane. Des exemples de filigrane complet ont été collectés sur d'autres documents.

- Filigrane

Hauteur × largeur maximales : 19,2 × 16,5 cm (relevées sur un document où le filigrane est complet).

Horloge octogonale, marquant 4 h moins 10 ; cadran, à partir du bord externe : bande subdivisée marquant les angles de l'octogone, suivie de 4 cercles concentriques et des heures en chiffres romains ; au-dessous, balancier dans un triple pentagone, la pointe vers le bas, suivi d'un double pentagone. Autour de l'horloge, 2 couronnes végétales réunies par un nœud placé dans la partie inférieure, dans l'axe vertical de la composition.

- Disposition du nom du marchand/revendeur et du filigrane sur le bifeuillet :

| 0,6 cm du 'N' de NUZZURALLY – 3,8 cm – Filigrane – 2,8 cm |

^ ^

21 cm 14 cm

v v

| 0,6 cm du 'N' de NUZZURALLY – 3,8 cm – Filigrane – 2,8 cm |

De nombreux autres exemples ont été identifiés par la contremarque ou le filigrane, mais ils sont de dimensions plus petites. L'un des documents rédigé sur la partie gauche de la feuille de papier où se trouve la contremarque, contient 4 dates : 20 mukarram 1891/1898, 27 mukarram 1891/1898, 1891/1899, 1891/1899. Un autre mentionne des soldats fuyant vers le Shewa. Un troisième parle d'al-Nūlā.

L'un des papiers à contremarque Nuzzerally Heptoolabhoy de Bombay a été utilisé pour la rédaction de codex en écriture *khojkī* : 2 viennent de la communauté ismaélienne de l'Inde de l'Ouest (Kutch et Gudjerat), estimés de 1896,

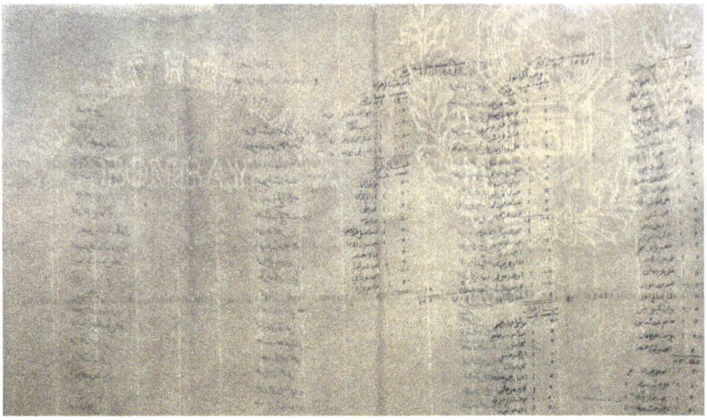

ILLUSTRATION 5.5A *Papier au nom de* NUZZURALLY HEBTOOLABHOY. *Bifeuillet complet*
© AIMABLE AUTORISATION DE L'IES, ADDIS ABEBA. PHOTOGRAPHIE ANNE REGOURD

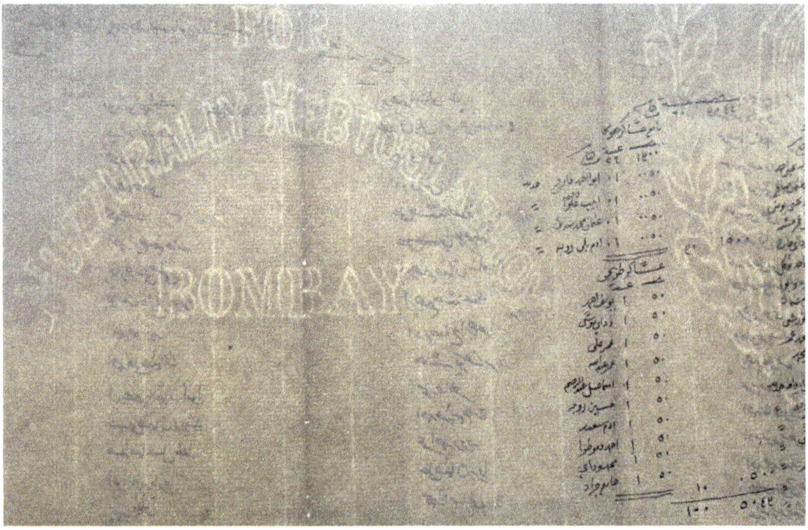

ILLUSTRATION 5.5B *Papier au nom de* NUZZURALLY HEBTOOLABHOY. *Contremarque*
© AIMABLE AUTORISATION DE L'IES, ADDIS ABEBA. PHOTOGRAPHIE ANNE REGOURD

ILLUSTRATION 5.5C
Papier au nom de NUZZURALLY HEBTOOLABHOY. *Filigrane*
© AIMABLE AUTORISATION DE L'IES, ADDIS ABEBA. PHOTOGRAPHIE ANNE REGOURD

présentés par Ali Sultaan Asani[11] ; 1 ms., conservé à Harvard, est estimé des alentours de 1890[12].

3.2.1.2.1 *Bombay* : MAHOMEDALLY [E]SSAJI

Document trouvé dans une partie du registre IES 963 non-paginée.

11 Il s'agit des mss. Ism K 2 et Ism K 12, dont le papier a été décrit par A. S. Asani, *Ecstasy and Enlightenment: The Ismaili Devotional Literature of South Asia*, 2002, 136, « Paper and Watermarks », chap. « Khojkī and its manuscript tradition ».

12 Ms. Indic 2534, Bibliothèque Houghton, Université de Harvard, numérisé dans le cadre du « Islamic Heritage Project », Harvard, « Sat:veṇī [niñḍhī] », avec « Dhrû:kathā [a story in 517 verses of Drū, a devotes of Kṛṣṇa] », en écriture Khojki, s. d., estimé des alentours de 1890, 20 × 16 cm, cf. Islamic Heritage Project, http://ocp.hul.harvard.edu/dl/ihp/011919175.

Texte
Date : 1890 al-muwāfiq 5 rabīʿ al-awwal 1316/24 juillet 1898 ; début : « ʿan bayānat ʿadad al-aṣnāf min al-ṣanādīq allatī fī dāḫilihā alwāḥ al-ruṣṣāṣ wa-rabṭāt silāḥ (…) », puis colonne.
Verso vierge.

Papier machine.
Dimensions maximales du document : 34 × 21 cm
Distance entre 2 fils de chaînette : 2,4 cm
20 vergeurs : 2,9 cm

- Contremarque

Hauteur de la lettre 'B' de BOMBAY : 1,8 cm ; hauteur du 'B' de BRITISH : 2,1 cm

BRITISH MAKE	(lettres doubles, rectiligne)
FOR	(lettres doubles, rectiligne)
MAHOMEDALLY [E]SSAJI	(lettres avec chevrons aux extrémités, semi-circulaire convexe)
BOMBAY	(lettres doubles, rectiligne, inscrit dans le demi-cercle du texte supérieur)

Distance entre 2 contremarques, du pied du 'M' de MAHOMEDALLY au sommet du 'B' de BRITISH : 20,3 cm.

Un autre document (17 × 10,4 cm), où figure la date de 18 al-qaʿda 1315 al-muwāfiq 3 (suivi d'un point ?) miāzīah 1890/10 avril 1898, est écrit sur un papier portant partiellement la même contremarque.

3.2.1.3 *Non localisé : BYRAMJI SHAPOORJI*
Document trouvé entre les pages 20 et 21 du registre IES 963.

Texte
Deux dates : 1891/1898–1899 ; début : « Ǧarrād Nūlī (?) … hīn qaṭaʿa lahu al-arḍ min qariyat al-Kābana (?) (…) ». Colonne de calcul.
Verso : « ʿadad », et 2 colonnes d'additions.

Papier machine. Apparence jaune clair.
Dimension maximale du document : 20,7 × 9,8 cm
Distance entre 2 fils de chaînette : 2,6 cm
20 vergeurs : 2,7 cm

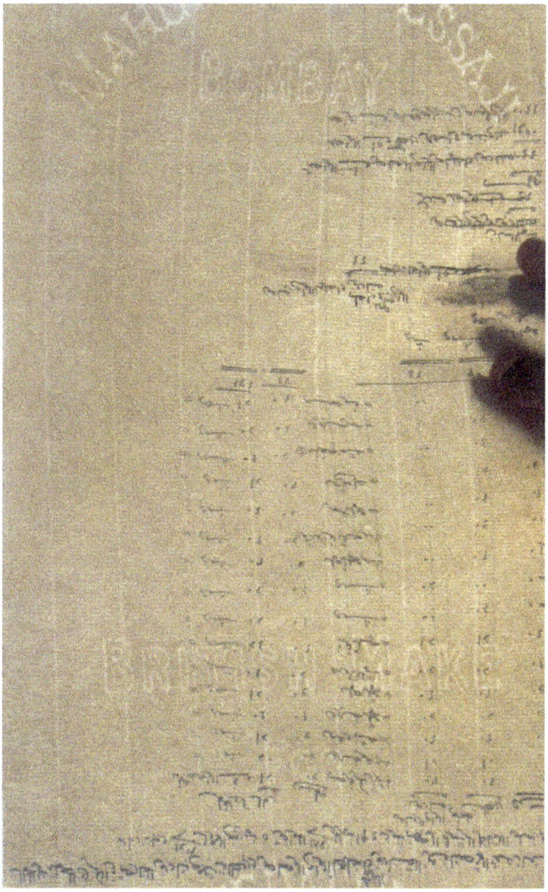

ILLUSTRATION 5.6A *Papier au nom de* MAHOMEDALLY [E]SSAJI © AIMABLE AUTORISATION DE L'IES, ADDIS ABEBA. PHOTOGRAPHIE ANNE REGOURD

ILLUSTRATION 5.6B *Papier au nom de* MAHOMEDALLY [E]SSAJI. *Vue partielle* © AIMABLE AUTORISATION DE L'IES, ADDIS ABEBA. PHOTOGRAPHIE ANNE REGOURD

ILLUSTRATION 5.6C *Papier au nom de* MAHOMEDALLY [E]SSAJI. *Vue partielle*
© AIMABLE AUTORISATION DE L'IES, ADDIS ABEBA. PHOTOGRAPHIE ANNE REGOURD

- Contremarque

Longueur maximale : 15,5 cm
Hauteur des lettres 'B' et 'S' : 1,7 cm

[BRITISH MAKE (?)] (lettres doubles, rectiligne)
BYRAMJI SHAPOORJI (lettres doubles, rectiligne)

Une première ligne appartenant à la contremarque se devine, d'autres ainsi qu'un filigrane ne sont pas exclus. Une seule occurrence de ce papier a été relevée dans le registre. L'un des manuscrits en sanscrit (*devanāgarī*), daté au colophon śaka 1803/1881, conservé par l'Université de Pennsylvanie, Rare Book & Manuscript Library, a été rédigé sur un papier au nom de Byramji Shapoorji[13].

13 Il s'agit du ms. Coll. 390, Item 543, contenant le Pañcaprakaraṇa de Rāmadāra, voir : http://dla.library.upenn.edu/dla/medren/record.html?id=MEDREN_5983973. Malheureusement, nous n'avons pas pu accéder à ce matériel.

ILLUSTRATION 5.7 *Papier au nom de* BYRAMJI SHAPOORJI

PHOTOGRAPHIE ANNE REGOURD

3.2.2 Papier fabriqué en Inde

3.2.2.1 *Calcutta : IMPERIAL PAPER MILLS*

Ramaseshan ne recense pour Calcutta que le Calcutta Paper Mill, créé vers 1789–1790 et le Royal paper Mill, créé en 1867[14].

Le document se trouvait entre les pages 75 et 76 du registre IES 963.

Texte
S. d. ; se déploie sur 8 colonnes dont deux doubles. L'entrée de la première colonne indique : « ʿan bayāna ʿan silāḥ allaḏī huwa bi-yad / ǧihat silāḥ min anfār ... ʿalā al-ṣandūq (...) »[15], la seconde et la troisième : « anfār ʿasākir ».
Verso vierge.

Papier machine.
Dimensions maximales du bifeuillet : 33,8 × 42 cm
Bords supérieurs et inférieurs tranchés nettement.

- Contremarque

Longueur maximale du texte : 14 cm
Entre les 2 lignes encadrant le texte : 6 cm

IMPERIAL PAPER MILLS C° L^{d}	(rectiligne)
CALCUTTA	(centré, rectiligne)

14 S. Ramaseshan, « The History of Paper in India up to 1948 », 1989, 116 et 117.
15 La barre oblique indique un changement de ligne.

point sous le 'o' et le 'd' qui sont en exposant.

- Filigrane

écu surmonté d'une couronne. Dans l'écu, initiales « IMP Co ». Et, en plus petit format, « L D » de part et d'autre du 'I'.

- Disposition du filigrane et du fabricant sur le bifeuillet

2,1 cm du bord supérieur		4,4 cm du bord supérieur	
^		^	
| 9,5 cm – Filigrane – 10,5 cm	–	Nom du fabricant – 2,2 cm |	(dans l'axe)
^		^	
19,5 cm du bord inférieur		23 cm du bord inférieur	

De nombreux autres documents volants ont été écrits sur ce papier :

- plusieurs d'entre eux, ayant la contremarque ou le filigrane ou les deux, ont des dates : 1890/1897 ; 1890/1897–1898 ; sur un autre document, dates autour de 1890/1897–1898 et 1891/1898–1899 et sur le dernier feuillet 11 mukarram 1891 muwāfiqihi 3 ǧumādā al-awwal 1316/19 septembre 1898 ; 1891/1898–1899 ; 1890/1898 et au verso 13 mukarram 1892/1899 ; 1891/1899 (deux fois sur le même document). Enfin, un document mentionne la présence du Ra's Makūnen à Harar (« ḥuḍūr rā's Makūnen ilā bandar Harar »[16]).
- outre Harar City, le Shewa est aussi mentionné, au sujet de soldats fuyards.
- un document a été réalisé dans un ternion de ce papier (dates autour de 1890–1891/1897–1899, avec, sur le dernier feuillet, un équivalent en années hégire : 11 mukarram 1891 muwāfiqihi 3 ǧumādā al-awwal 1316/19 septembre 1898). Les dimensions maximales d'un feuillet sont de 21 × 16,7 cm ; d'un bifeuillet à l'autre, on a une alternance filigrane/contremaque : a. bifeuillet externe : filigrane ; b. bifeuillet médian : contremarque ; c. bifeuillet central : filigrane.

Un document sur papier à en-tête « Imperial Paper Mills Company » de Calcutta, datant du 15 février 1892, est accessible en ligne[17].

16 Le linguiste ne manquera pas de noter la présence du mot « Bandar » pour « ville », présent en Égypte.

17 http://www.hwph.de/historische-waardepapieren/losnr-auktnr-pa17-44_1.html.

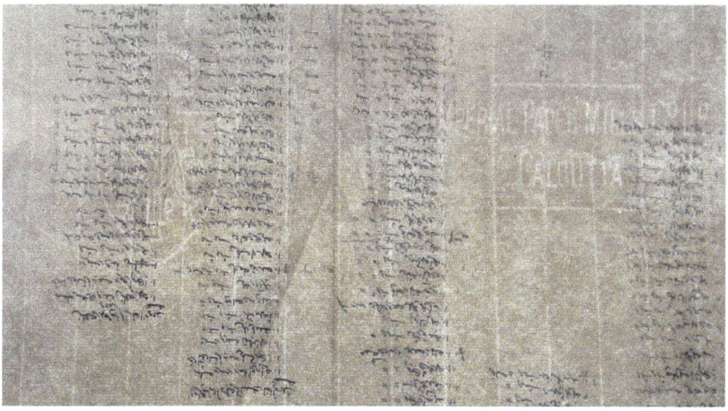

ILLUSTRATION 5.8A *Papier* IMPERIAL PAPER MILLS C°L^{d}
© AIMABLE AUTORISATION DE L'IES, ADDIS ABEBA. PHOTOGRAPHIE ANNE REGOURD

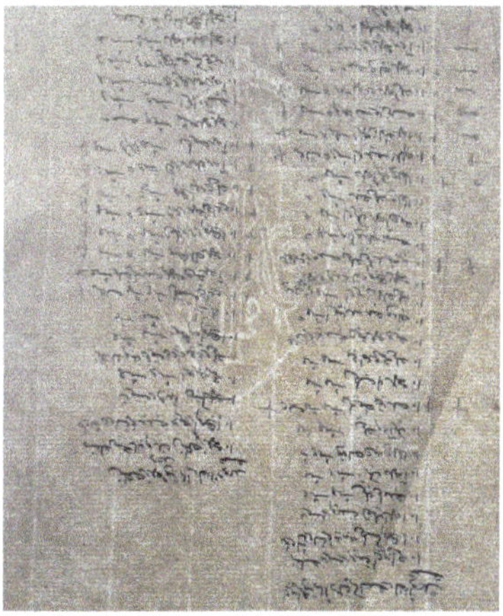

ILLUSTRATION 5.8B
Papier IMPERIAL PAPER MILLS C°L^{d}*. Filigrane*
© AIMABLE AUTORISATION DE L'IES, ADDIS ABEBA. PHOTOGRAPHIE ANNE REGOURD

3.2.3 Fabrication non-identifiée

3.2.3.1 *Bombay ? : N* CUMRO[

Texte
S. d. ; début : « ʿasākir al-mutawağğihīn / ʿadad ».
Verso : 2 additions + « qāʾil ʿan Muḥammad wa-... ḍamānat Muḥammad (...) ».

Papier machine. Jaune clair.
Dimensions maximales du document : 22,6 × 10,5 cm

ILLUSTRATION 5.8C *Papier IMPERIAL PAPER MILLS C° L^d^.*
Contremarque

Distance entre 2 fils de chaînette : 2,5 cm
20 vergeurs : 2,9 cm

- Contremarque

N CUMRO[
    BOM[BAY ?

Dans le Phillips' Paper Trade Directory of the World, parmi les « Stationers » à Bombay, apparaît un : Cumrodin, Abdulrahim & Bros., Parsee Bazaar Street[18].

3.2.3.2 *Papier « KAISAR I HIND »*
Le document a été trouvé entre les pages 67 et 68 du registre IES 963. Il a été plié en 2 verticalement et en 4 horizontalement.

Texte
dates : 1890 al-muwāfiq rabīʿ al-awwal 1316/juillet-août 1898 ; début : « ʿan bayānat ʿadad al-aṣnāf min al-ṣanādiq allatī fī dāḫilihā alwāḥ al-ruṣṣāṣ (…) ». En colonne, une liste de noms.

---

18 *Phillips Paper Trade Directory of the World* 1923, 527.

Verso : horizontalement et tête-bêche, des noms et des calculs.

Papier machine. Très fin. Jaune clair.
Dimensions maximales du document : 34,4 × 21 cm
Distance entre deux fils de chaînette : 2,5 cm
20 vergeurs : 2,8 cm

- Contremarque

longueur 1re l. contremarque (du K au D) : 13 cm ; hauteur lettre 'K' (plus haute) : 2,1 cm
longueur 2de l. contremarque : 9,7 cm ; hauteur lettre 'F' (plus haute) : 1,7 cm

KAISAR I HIND	(lettres doubles, semi-circulaire convexe)
FOOLSCAP	(lettres doubles, rectiligne)

« Foolscap » : souvent traduit par « papier écolier », désigne un papier coupé à la taille de 8 ½ × 13 ½ inches (34,3 × 21,6 cm), quelquefois 13 × 8 inches (33 × 20 cm). Les dimensions de notre document correspondent aux dimensions d'une feuille dans ce format.

- Filigrane

Placé entre la 1re et la 2de l. de la contremarque
hauteur × largeur : 7,5 × 6,4 cm

médaillon ovale formé d'un trait simple, contenant le portrait d'une reine en buste, de profil, tournée vers le gauche. Il s'agit certainement de la Reine Victoria (r. 20 juin 1837 à sa mort, le 22 janvier 1901) : son profil et sa couronne sont reconnaissables, mais, de plus, en 1877, la Reine a été proclamée « Empress of India », Impératrice d'Inde, soit, en hindi, *Kaisar-i-Hind* ; son image apparaît alors sur divers supports, monnaies, timbres-poste, portraits, et, on le voit, filigranes du papier. En 1900, une médaille est créée, la « Kaisar-I-Hind Medal », médaille Kaisar-I-Hind : il s'agit d'une distinction civile britannique suivant le nom de l'Empereur indien Kaisar-I-Hind, récompensant les services rendus à l'Inde britannique.

Quatre autres documents, aux dimensions plus petites, sont écrits sur ce papier, l'un comporte une date en 1889/1896.

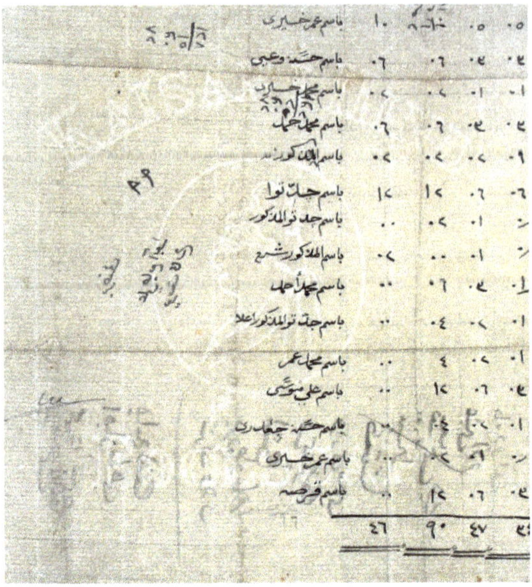

ILLUSTRATION 5.9 *Papier* KAISAR I HIND
© AIMABLE AUTORISATION DE L'IES, ADDIS ABEBA.
PHOTOGRAPHIE ANNE REGOURD

4 Papiers additionnels de codex du pays Oromo

Des papiers similaires ont été trouvés dans le pays Oromo. Les manuscrits dont ils proviennent ne sont pas datés et, pour ceux de Tije et de Suuse, ils correspondent à des feuillets libres. Cependant, la série relevée jusque-là faisant corpus, ces papiers peuvent faire l'objet d'une première localisation dans l'espace et le temps, compte tenu de leurs marques. Quatre d'entre eux sur 6 mentionnent Bombay et une fabrication britannique (« British make »), surtout, les noms, précieux, qu'ils portent, ajoutent à la première liste.

4.1 *Limmū Gannat (Limmu Genet / Suntu), Jimma*

4.1.1 LMSG00005; LMSG00049[19]

Papier machine.

19 Lien à la base de données d'« Islam in the Horn of Africa », GLG00005, http://islhornafr.tors.sc.ku.dk/backend/manuscripts/597; LGLG00049, http://islhornafr.tors.sc.ku.dk/backend/manuscripts/735.

ILLUSTRATION 5.10 *Ms. LMSG00005. Papier* M. A. MIA SAHEB SHURUFALLY *avec navire. Marques complètes. La partie supérieure des mâts du navire est visible dans la partie inférieure du document*

- Contremarque

M.A. MIA SAHEB SHURUFALLY	(lettres doubles, semi-circulaire convexe)
BRITISH	(centré, rectiligne)
MAKE	(centré, rectiligne)
MANO WAR TRADE MARK	(rectiligne)
BOMBAY	(lettres doubles, semi-circulaire centré)

- Filigrane

Placé entre la 1re et la 2de l. de la contremarque

... ? Navire à voile sur l'eau ... ?[20] (rectiligne)

20 Marques à droite et à gauche non-déchiffrées. À droite, on note une lettre ressemblant à un *qāf*.

ILLUSTRATION 5.11 *Ms. LMSG00049. Papier* M. A. MIA SAHEB SHURUFALLY
© PHOTOGRAPHIE SARA FANI

4.1.2 LMSG00049

Papier machine.

M. A. MIA SAHEB SHURUFALLY	(lettres doubles, convexe, volutes aux initiales)
BOMBAY	(lettres doubles, centré, rectiligne, volute lettre initiale)

Volute après Bombay.

4.1.3 LMSG00005 ; LMSG00044[21] ; LMSG00049

C'est à l'emblème du lion de Judah que Menelik II s'est référé – et non au motif persan du lion au soleil, brandissant un sabre, que l'on trouve représenté dans ce papier.

21 LGLG00044, http://islhornafr.tors.sc.ku.dk/backend/manuscripts/719.

Papier machine.

- Contremarque

Āftāb ʿĀlimkīr (*kāf* persan, گ)	(lettres doubles, centré, rectiligne)
SILVER SHADE	(centré, rectiligne)
LION BRAND	(centré, rectiligne)
SAMIER BE-NAZEER	(lettres cursives doubles, centré, rectiligne)
BRITISH MAKE	(lettres doubles, centré, rectiligne)

- Filigrane

Placé entre les 2 1res et les 3 dernières l. de la contremarque

Soleil
Lion brandissant un sabre

4.1.4 LMSG00005 ; LMSG00044 ; LMSG00049

Papier machine.

- Contremarque

SAMIER BE NAZEER	(caractères persans, urdus ?)
BRITISH MAKE	(lettres doubles, centré, rectiligne)

- Filigrane

Placé entre la 1re et la 2de l. de la contremarque

Soleil
Écu

4.2 *Tije, Jimma*

Le site a fait partie de l'ancien royaume de Gomma. La collection de manuscrits est liée à la mosquée et au mausolée de Shekota Tije al-Qurārī (m. 1917 ?), un maître de la Confrérie de la Tiǧāniyya qui épousa une fille d'Abba Jifar II, Roi du royaume de Gibe, Jimma (r. 1878–1932). Ces manuscrits, placés à l'extérieur de la mosquée et du mausolée, sont donc dans un piètre état et consistent en fragments, feuillets volants ou cahiers désorganisés.

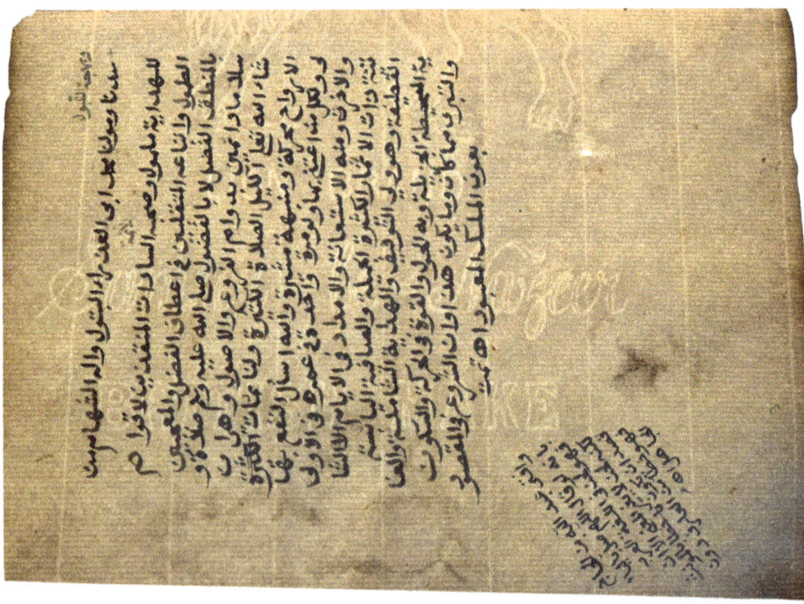

ILLUSTRATION 5.12A, B *Ms. LMSG00005. Papier* SAMIER BE-NAZEER *au lion et soleil. Marques, parties supérieure et inférieure*
© PHOTOGRAPHIE SARA FANI

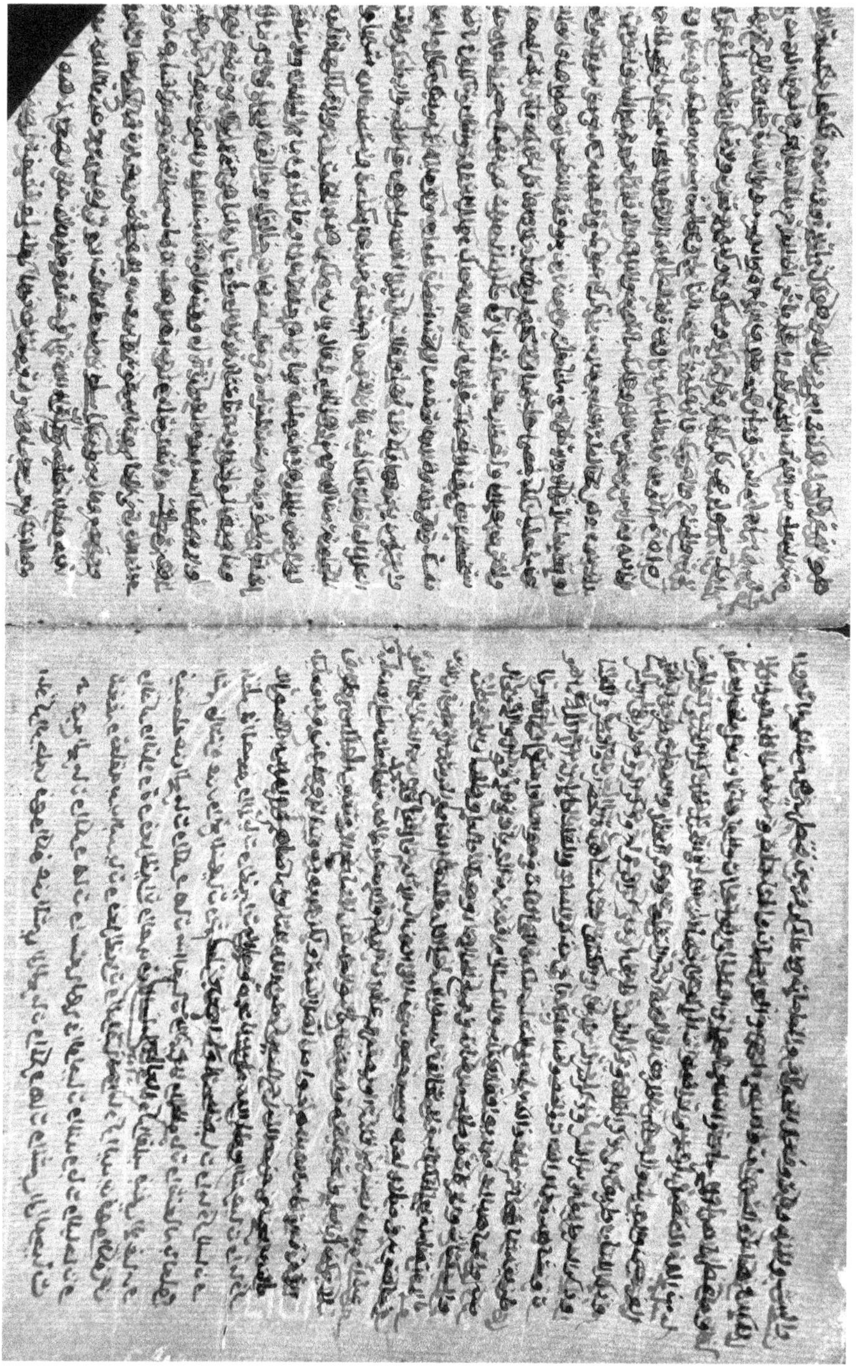

ILLUSTRATION 5.12C *Ms. LMSG00044. Même papier. Marques quasiment complètes*
© PHOTOGRAPHIE SARA FANI

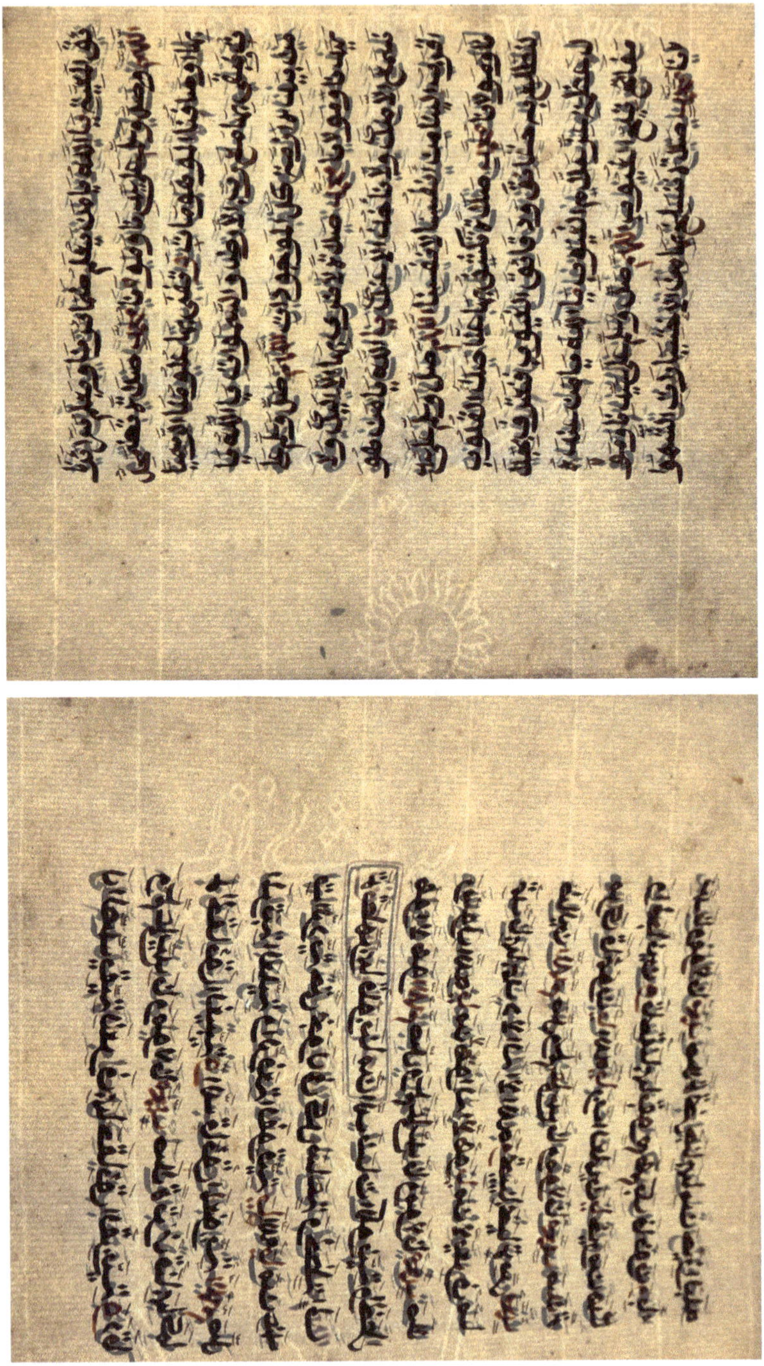

ILLUSTRATION 5.13A, B *Ms. LMSG00049. Papier* SAMIER BE-NAZEER *à écu et soleil. Marques supérieure et inférieure*
© PHOTOGRAPHIE SARA FANI

ILLUSTRATION 5.14 *Ms. Tije, feuillet libre. Papier* HUSSUNALLY ABDOOLA[
© PHOTOGRAPHIE SARA FANI

Feuillet libre

Papier machine.

BOMBAY STAT ... A	(lettres doubles, concave)
BRITISH MAKE	(lettres doubles, centré, rectiligne)
HUSSUNALLY ABDOOLA[	(lettres doubles, centré, rectiligne)

4.3 *Suuse*[22]

SUSU00027, f. 21r–v (feuillet libre)

Papier machine.

IASAHEB (ASAHEB ?) (cursives, nombreuses boucles angulaires, oblique)
ABDOOLALLY
IASAHEB (ASAHEB ?)
SHURAFALLY
BOMBAY

22 SUSU00027, http://islhornafr.tors.sc.ku.dk/backend/manuscripts/1501.

ILLUSTRATION 5.15 *Ms. SUSU00027, f. 21r–v (feuillet libre). Papier IASAHEB (ASAHEB ?) ABDOOLALLY IASAHEB (ASAHEB ?) SHURAFALLY*
© PHOTOGRAPHIE SARA FANI

5 Discussion & conclusion

Les documents réunis ici datent d'entre 1893 et 1918–1919, mais la majorité se rapporte aux années 1898–1899 ; la contremarque de l'un des papiers contient la date de 1899 et le filigrane du papier à l'effigie de la Reine Victoria, « Impératrice de l'Inde », le place entre 1877 et sa mort en janvier 1901 – à moins qu'il n'ait été conservé à titre commémoratif.

La plupart des papiers sont aux marques de marchands de la côte occidentale de l'Inde, principalement de Bombay, l'un d'eux a des connexions avec Ahmedabad, dans le Gudjerat – seul le papier sortant de moulins locaux provient de la côte est, de Calcutta, sur le Golfe du Bengale. Les papiers sont faits sur commande, en Angleterre surtout, mais aussi en Belgique, lorsque c'est indiqué. Les contremarques sont en caractères latins, mais l'une est en alphabet arabe/persan : les marques omniprésentes, couvrant littéralement la surface du papier, sont une autre caractéristique du papier MAHOMEDALLY[BHOY ?] WALLY IARVARI produit en Angleterre, qu'il partage avec le papier de NUZZURALLY HEBTOOLABHOY au filigrane à horloge et avec le papier d'ALLIBHOY AKBERALLI de Bombay, Sutar Shawl, à filigrane à l'effigie d'Édouard VII et de son épouse, fabriqué en Angleterre et trouvé au Yémen[23]. Ils sont produits pour un marché donné, en étroite collaboration avec des manufactures occidentales. Il s'agit là d'un début de corpus, mais qui documente les circuits commerciaux et soulève la question des routes du trafic entre l'Inde et l'Éthiopie.

Dans les dernières décennies du XIXe s. et les premières années du XXe s., le commerce est à peu près entièrement entre les mains des étrangers en Éthiopie : parmi eux, les Indiens sont en bonne place[24]. Notre corpus se place à cette période, charnière par la naissance d'une capitale, les événements au Harar et l'expansion des intérêts indiens[25]. Le développement d'Addis Abeba à la fin du XIXe s., suivant sa fondation en 1886, attire, sur l'impulsion de Menelik II, des firmes basées à Aden, outre des artisans spécialisés dans la construction[26]. Parmi les plus importantes, on note celle de Menahim Messa, connu sous le nom de Benin : fondée en 1850, elle était spécialisée dans

23 A. Regourd, « Papiers filigranés de manuscrits de Zabīd, premier tiers du XVIIIe-milieu du XXe siècle : papiers importés et 'locaux' », 2015, 238*sq*.

24 R. K. P. Pankhurst, « Menilek and the Utilisation of Foreign Skills in Ethiopia », janvier 1967, 75. Au sujet du moment d'inversion de la domination française sur le commerce d'import-export, en 1896, au profit des Indiens et des Arméniens, cf. P. P. Garritson, *A History of Addis Abäba from its Foundation in 1886 to 1910*, 2000, 16–17.

25 Sur la division en deux périodes dans la démographie de l'émigration indienne dans l'océan Indien, avec 1880 comme date charnière, voir : C. Markovits, « Indian Merchant Networks outside India in the Nineteenth and Twentieth Centuries: A Preliminary Survey », 1999, 891*sq*.

26 S. A. Bezabeh, *Subjects of Empires/Citizens of States: Yemenis in Djibouti and Ethiopia*, 2016.

l'exportation d'ivoire, de café et de peaux et l'importation de textiles et de fournitures domestiques[27]. La même politique est poursuivie à Harar, qui passe sous le contrôle du Ra's Mekonnen Welde Mikaél (1852–1906), nommé par Menelik Gouverneur du Hararghe en 1887, privilégiant en un second temps, aux côtés des Français, l'installation d'Indiens de Bombay basés à Aden, aux côtés d'Arabes du même port, maîtrisant des métiers divers : suivant l'offensive italienne de 1895, la plus importante firme commerciale indienne, celle de G. M. Mohamedally (1886–1937), fondée par un groupe de marchands Bohras dāwūdites de Bombay avec l'appui du Ra's Mekonnen, s'établit à Harar City ; elle est destinée à jouer un rôle majeur dans le développement du commerce éthiopien[28]. Mais c'est l'ouverture des relations anglo-éthiopiennes, en 1897, qui est responsable d'un afflux notable de marchands et artisans indiens, sujets britanniques, en Éthiopie, un afflux dont la jeune capitale bénéficie, à côté de l'apport de firmes déjà établies dans le Harar[29].

Leur approvisionnement à proximité, à Aden ou à Bombay, garantissant des coûts de transport réduits, donc des prix concurrentiels, a été souligné comme l'une des clés de leur succès, auquel s'ajoute leur connaissance du marché et des goûts des consommateurs : nombre d'entre eux ont l'expérience du commerce

27 R. K. P. Pankhurst, A. Amba & Adjacent Markets, « The Trade of Central Ethiopia in the Nineteenth and Early Twentieth Centuries », juillet 1964, 55 ; R. K. P. Pankhurst, « Menilek and the Utilisation of Foreign Skills in Ethiopia », janvier 1967, 75–76 ; *id.*, « Across the Red Sea and Gulf of Aden: Ethiopia's Historic Ties with Yaman », 2002, 417. R. Ahroni, *The Jews of the British Crown Colony of Aden. History, Culture, and Ethnic Relations*, 1994, 46*sq.*

28 R. K. P. Pankhurst, « Menilek and the Utilisation of Foreign Skills in Ethiopia », janvier 1967, 37, 51 ; R. K. P. Pankhurst, A. Amba & Adjacent Markets, « The Trade of Central Ethiopia in the Nineteenth and Early Twentieth Centuries », juillet 1964, 76. Au sujet des marchands indiens avant l'arrivée de Menelik à Harar, voir le cas de Haji Yusuf Barkatly, qui, d'après Bardey, l'employeur d'Arthur Rimbaud basé à Aden, était le plus important vendeur d'esclave de la ville ; il s'enfuit à Zayla après que les Éthiopiens ont mis fin à l'occupation égyptienne, *id.*, juillet 1964, 77. Sur la « filière » Bombay-Aden-Corne de l'Afrique, P. P. Garritson, *A History of Addis Abäba from its Foundation in 1886 to 1910*, 2000, 65. Voir la monographie que D. Harre a consacré à G. M. Mohamedally, « The Indian Firm G. M. Mohamedally & Co in Ethiopia (1886–1937) », 2015, 290.

29 R. K. P. Pankhurst, « Menilek and the Utilisation of Foreign Skills in Ethiopia », janvier 1967, 55. R. K. P. Pankhurst, A. Amba & Adjacent Markets, « The Trade of Central Ethiopia in the Nineteenth and Early Twentieth Centuries », juillet 1964, 54. Le même phénomène de présence avant le Traité avec les Anglais, puis d'afflux avec le Traité, est constaté pour les artisans travaillant dans la construction, voir R. K. P. Pankhurst, « The role of Indian craftsmen in late nineteenth and early twentieth-century Ethiopian palace, church and other building », 1995, 13–14 ; P. P. Garritson, *A History of Addis Abäba from its Foundation in 1886 to 1910*, 2000, 66, 127–128.

en Afrique ou au Moyen-Orient[30]. Mais les Indiens bénéficient aussi de l'accès à la poste hebdomadaire, ouverte à tout sujet britannique[31].

Les deux firmes indiennes les plus importantes de Harar sont celles de Mohamedally et de Taib Ali Akbarali, tous deux originaires de Bombay. Ces grandes compagnies d'import-export sont précisément celles qu'il faut suivre, en ce qu'elles dominent ou contrôlent les unités opérant à petite échelle de plusieurs façons[32]. En 1905, poursuivant son expansion, Mohamedally installe à Anderacha (Kaffa), dont le marché a grandi en importance au début du xxe s. à la faveur de l'arrivée de marchands étrangers, une branche dirigée par un Indien, Abdul Hussein, qui s'occupait principalement de l'exportation de café et de l'importation de divers biens manufacturés[33]. Il hérite du monopole du tabac éthiopien, détenu par la Régie Co-intéressée des Tabacs de l'Empire Éthiopie, entre 1908 et 1909[34]. Plus exceptionnellement, G. M. Mohamedally se met aussi à produire du savon autour de 1905[35]. Au tournant du siècle, le commerce indien s'impose, à Harar de même qu'à Addis Abeba, secouant les intérêts français[36]. Il semble qu'on assiste ensuite à un redéploiement des Indiens d'Addis vers l'intérieur du pays : ils s'installent, entre autre, à Jimma, Ghinnir, Lekemti, Däbrä Marqos, Bali et Walamo, contribuant ainsi à la structuration

30 H. Le Roux, *Menelik et nous : Le Carrefour d'Aden*, 1902, 146. R. K. P. Pankhurst, A. Amba & Adjacent Markets, « The Trade of Central Ethiopia in the Nineteenth and Early Twentieth Centuries », juillet 1964, 54 ; R. K. P. Pankhurst, « Menilek and the Utilisation of Foreign Skills in Ethiopia », janvier 1967, 54.

31 R. K. P. Pankhurst, A. Amba & Adjacent Markets, « The Trade of Central Ethiopia in the Nineteenth and Early Twentieth Centuries », juillet 1964, 54. D. Harre, « The Indian Firm G. M. Mohamedally & Co in Ethiopia (1886–1937) », 2015, 297, relève l'usage des routes du Soudan égyptien accordé par les Anglais à la firme au moins en 1908.

32 P. P. Garritson, *A History of Addis Abäba from its Foundation in 1886 to 1910*, 2000, 65, 118.

33 P. P. Garritson, *A History of Addis Abäba from its Foundation in 1886 to 1910*, 2000. R. K. P. Pankhurst, « The Trade of Southern and Western Ethiopia and the Indian Ocean Ports in the Nineteenth and Early Twentieth Centuries », juillet 1965, 65. D. Harre, « The Indian Firm G. M. Mohamedally & Co in Ethiopia (1886–1937) », 2015, 297.

34 R. K. P. Pankhurst, « Menilek and the Utilisation of Foreign Skills in Ethiopia », janvier 1967, 70.

35 R. K. P. Pankhurst, « Menilek and the Utilisation of Foreign Skills in Ethiopia », janvier 1967, 73. A. Zervos, *L'empire d'Éthiopie*, 1936, 145, atteste dans les années 1930 de la production d'autres produits encore. D. Harre, « The Indian Firm G. M. Mohamedally & Co in Ethiopia (1886–1937) », 2015, 301.

36 H. Leroux, *Menelik et nous : Le Carrefour d'Aden*, 1902, 146, dit, à la mi-janvier 1901, que « le commerce de Harar est accaparé par ces deux maisons indiennes ». R. K. P. Pankhurst, « The Trade of Southern and Western Ethiopia and the Indian Ocean Ports in the Nineteenth and Early Twentieth Centuries », juillet 1965, 57.

du marché éthiopien à partir de la capitale[37]. Au milieu des années 1930, on ne compte pas moins de 50 firmes ayant des branches à Jimma, ainsi que 97 étrangers, dont 23 Indiens et 21 Arabes, résidant dans la province[38]. Ceci sans présumer de la circulation elle-même des produits par le biais des nombreux marchés[39].

Au début du xxe s., un rapport britannique donne une liste des marchands les plus considérables d'Addis, par nationalité : parmi les plus grands de l'import-export, figurent les Indiens, dont au premier rang, Mohamedally et son associé, Golam Ali, Djewadji et Yusuf Ali[40]. En 1907, parmi les principaux marchands en Éthiopie, on compte trois compagnies indiennes : Golam Ali Mohamed Ali and Co (import-export), Djewadji (import-export) et Akbar Ali and Co.[41]. Un rapport de 1913 indique que le commerce dans la province éthiopienne de Harar est entre les mains de Grecs, d'Arméniens, mais surtout d'Indiens : parmi les produits importés, figure le papier[42].

---

37 R. K. P. Pankhurst, A. Amba & Adjacent Markets, « The Trade of Central Ethiopia in the Nineteenth and Early Twentieth Centuries », juillet 1964, 54. Voir aussi R. K. P. Pankhurst, « Menilek and the Utilisation of Foreign Skills in Ethiopia », janvier 1967, 77 : « Indian traders also penetrated fairly widely. Thus Faitlovitch reported their presence in 1908–9 in the Eritrean towns Asmara, Adi Caieh and Saganeiti, while Boyes a few years earlier had met one near Lake Margherita ». La présence indienne dans un certain nombre de places précède la vague qui fait l'objet de notre étude, au tournant du siècle, le port de Massawa avec ses liens à Aden et Bombay, en est un bon exemple, voir R. K. P. Pankhurst, « The Trade of Northern Ethiopia in the Nineteenth and early Twentieth Centuries », janvier 1964, 93–94. D. Harre, « The Indian Firm G. M. Mohamedally & Co in Ethiopia (1886–1937) », 2015, 300, branches de la Cie progressivement ouvertes à Dire Dawa, Jijiga, Jimma, Lekempté, Modjo, Wellamo. On observe un phénomène similaire pour le commerce arménien à partir des années 1930, voir B. Adjemian, *La fanfare du Négus : les Arméniens en Éthiopie (XIXe–XXe siècles)*, 2013, 241.

38 R. K. P. Pankhurst, « The Trade of Southern and Western Ethiopia and the Indian Ocean Ports in the Nineteenth and Early Twentieth Centuries », juillet 1965, 59.

39 R. K. P. Pankhurst, « The Trade of Northern Ethiopia in the Nineteenth and early Twentieth Centuries », janvier 1964, selon Zervos, 69 ; voir différentes villes abordées dans l'article.

40 R. K. P. Pankhurst, « The Trade of Southern and Western Ethiopia and the Indian Ocean Ports in the Nineteenth and Early Twentieth Centuries », juillet 1965, 75–76.

41 R. K. P. Pankhurst, A. Amba & Adjacent Markets, « The Trade of Central Ethiopia in the Nineteenth and Early Twentieth Centuries », juillet 1964, 55.

42 E. Wagner, *Harar. Annotierte Bibliographie zum Schrifttum über die Stadt und den Islam in Südostäthiopien*, 2003, notice 47.16, 494. Pour les Arméniens, voir B. Adjemian, *La fanfare du Négus : les Arméniens en Éthiopie (XIXe–XXe siècles)*, 2013, 239*sq.*, dont 240, où le papier apparaît parmi les articles vendus par les détaillants arméniens (« petit commerce »), qui représentaient la plupart des commerçants en 1908, pour satisfaire la demande des étrangers et des Éthiopiens.

Cependant, rapports et statistiques demeurent flous lorsqu'ils se réfèrent à des nationalités, des ethnies ou des communautés socio-religieuses : « les Indiens », « les Arabes », … Ils conduisent à sous-estimer les connexions avec Aden et sa place dans le circuit commercial, de même que l'importance des réseaux. À Aden, colonie de la Courone anglaise entre 1839 et 1967, contrôlée depuis Bombay jusqu'en 1937, c'est sur le tandem anglo-indien (Hindous, dont Banyans) que repose la machine économique et légale[43]. Massawa, de même qu'Aden, compte de nombreux Hindous (dont des Banyans), il y a des Parsis et, pour Aden et Harar, des Bohras[44]. La présence de Khojas est attestée par ailleurs[45]. Les marchands indiens exportateurs de papier attestés par le corpus présenté ici sont en bonne partie des musulmans de Bombay. Le suffixe BHOY suggère qu'il s'agit d'Ismaéliens, sans plus de précision.

43 Sir K. Trevaskis, *Shades of Amber: A South Arabian Episode*, 1968, 4, au sujet d'Aden « an Anglo-Indian colony, where it was largely British and Indian hands that made the wheels go round ». Pour une vision contrastée du modèle des marchands indiens à l'étranger, occupant une position intermédiaire ou subalterne (« middlemen ») sur la période, voir C. Markovits, « Indian Merchant Networks outside India in the Nineteenth and Twentieth Centuries: A Preliminary Survey », 1999, 909*sq*.

44 R. K. P. Pankhurst, « The Trade of Northern Ethiopia in the Nineteenth and early Twentieth Centuries », janvier 1964, 92, « Rassam (…) adds, however, that the Banian merchants, being Hindu, objected to drinking water brought in animal skins and therefore subsisted mainly on rainwater » ; *ibid.*, 96 ; Stafford B. Haines mentionne 50 Baniyans lors de la prise d'Aden par les Anglais en 1839, « Memoirs on the South and East Coast of Arabia », 1845 ; pour les années suivant l'occupation anglaise, Z. H. Kour, *The History of Aden*, 2004, 21–22 ; R. Ahroni, *The Jews of the British Crown Colony of Aden. History, Culture, and Ethnic Relations*, 1994, 42 ; D. Harre, « The Indian Firm G. M. Mohamedally & Co in Ethiopia (1886–1937) », 2015, 288*sq.*, en particulier 290, 298–299.

45 Voir le tableau 2 détaillant par caste/destination aux XIXe et XXe s. et les commentaires de C. Markovits, « Indian Merchant Networks outside India in the Nineteenth and Twentieth Centuries: A Preliminary Survey », 1999, 898. B. Adjemian, *La fanfare du Négus : les Arméniens en Éthiopie (XIXe–XXe siècles)*, 2013, 59*sq*. Au Kenya, en 2008, des Indo-Kenyans étaient présents dans l'industrie de la papéterie et de l'emballage (carton), dont des Jaïns Oshwal, des hindous et des Ismaéliens (Khojas), et étaient à l'origine d'une part importante du chiffre d'affaire et de l'emploi dans le pays, attestant parfois d'investissements de l'Agha Khan ; le portrait de l'Ismaélien Sultan D. donne l'exemple d'un chef d'entreprise à la tête d'une fabrique de papier à partir du recyclage des journaux et des vieux papiers, créée en 1990, voir M. Adam, « Des Indiens des comptoirs aux Indo-Africains », *id.*, « Panorama des communautés socio-religieuses », et *id.*, « Portraits et fragments d'histoires de vie au Kenya », 46, 56, 97 et note 39, et G. G. Wairire, « Commerçants et industriels d'origine indo-pakistanaise au Kenya. Regards sociologiques », 279, 280, 282, dans : M. Adam (éd.), *L'Afrique indienne. Les minorités d'origine indo-pakistanaise en Afrique orientale*, 2010, 425.

G. M. Mohamedally, ce riche et talentueux marchand Bohra dāwūdite, qui prend son essor de Harar City, est peut-être le commanditaire du papier à contremarque *MAHOMEDALLYBHOY* par le biais de sa parentèle de Bombay[46]. Le papier au nom du marchand ALLIBHOY AKBERALLI, lui aussi de Bombay, quartier de Sutar Shawl, fabriqué en Angleterre et ayant servi à la copie d'un livre manuscrit au Yémen, atteste des mêmes réseaux : trouvé à Zabid, il a certainement touché terre dans le port d'Aden[47]. Les documents réunis ici sont visiblement administratifs, mais en Éthiopie, le papier est typiquement utilisé pour la production de manuscrits islamiques[48] et Harar en est l'un des centres les plus importants dans le pays, représentant un marché potentiel certain, distinct de celui formé par les consommateurs européens. D'après ce que l'on sait à ce jour, les mêmes papiers de ces marchands de Bombay ont été utilisés pour des livres manuscrits par les Ismaéliens de Kutch et du Gudjerat, sans doute par des Bohras dāwūdites (collections privées)[49]. Les marchands avaient aussi besoin de papier pour leurs calculs, leurs comptes ou leur correspondance.

Plus étrangement, l'autre imprécision des rapports et statistiques porte sur le poste « papiers ». Il s'agit d'articles importés, indispensables à la

46 Selon D. Harre, si le fondateur de la firme est nommé et connu par le nom de la firme qu'il a fondée, il faut rester prudent sur les identifications du fait de personnes portant le même nom, à l'intérieur-même de la firme, voir « The Indian Firm G. M. Mohamedally & Co in Ethiopia (1886–1937) », 2015, dans son chap. 1, 286–288.

47 A. Regourd, « Papiers filigranés de manuscrits de Zabīd, premier tiers du XVIIIe-milieu du XXe siècle : papiers importés et 'locaux' », 2015, 238*sq*., spécialement d., 240. Sur la segmentation des réseaux (networks) indiens et les niches économiques occupées par chacun, voir C. Markovits, « Indian Merchant Networks outside India in the Nineteenth and Twentieth Centuries: A Preliminary Survey », 1999, 902–903.

48 A. Regourd, « Introduction to the Codicology: Papers, Ruling, and Bindings », 2014, xlix. R. K. P. Pankhurst, « The Trade of Northern Ethiopia in the Nineteenth and early Twentieth Centuries », janvier 1964, 73, signale l'utilisation de papier importé par des églises et monastères dès le xixe s., mais en quantité limitée, indice de contact avec les Occidentaux, « The growing contact with foreign lands may be illustrated by a certain limited use in the nineteenth century of imported paper in some of the Ethiopian churches and monasteries ».

49 O. Akkerman, The Bohra Dark Archive and the Language of Secrecy: A Codicological Ethnography of the Royal ʿAlawī Bohra Library in Baroda, juillet 2015, 250–251 et, surtout, ill. 87, 253, montre un papier EXTRA SUPERFINE, suivi d'un texte en arabe et de clés croisées présentés comme "Bohra Watermarks", sans autre indication. Les papiers européens sont décrits comme venant d'Angleterre pour nombre d'entre eux ; des formats *Foolscap Universal* sont signalés. Ils sont davantage présents dans les collections privées par opposition aux « Dark Archives ».

documentation de l'activité douanière, commerciale et aux échanges « diplomatiques » entre compagnies et gouvernants, pour ne se borner qu'à ces secteurs[50]. On a une quantité et un montant pour Massawa au milieu du xixᵉ s., où figurent, parmi les importations en provenance d'Arabie et d'Égypte, 12,712 kg de papier pour une valeur de 300 Maria Theresa dollars[51]. La provenance, ainsi que les papiers classés selon leur usage, n'apparaissent que rarement : on a relevé deux fois « writing paper »[52], et, parmi les articles importé à Kaffa en 1905, du papier à écrire d'Autriche et de France, dont une partie est transportée via l'Inde : l'auteur ajoute, qu'à Anderacha une feuille de papier coûtait une barre de sel, tandis qu'à Sharada, on avait 4 feuilles seulement pour 1 dollar ou 80 feuilles pour 20 dollars[53].

Bibliographie

Akkerman, Olly. Juillet 2015. The Bohra Dark Archive and the Language of Secrecy: A Codicological Ethnography of the Royal ʿAlawī Bohra Library in Baroda, Ph.D. Dissertation, Université libre de Berlin.

Adam, Michel. 2010. « Des Indiens des comptoirs aux Indo-Africains », dans : M. Adam (éd.), *L'Afrique indienne. Les minorités d'origine indo-pakistanaise en Afrique orientale*, Paris, IFRA/Karthala, 15–78.

Adam, Michel. « Panorama des communautés socio-religieuses », dans : M. Adam (éd.), *L'Afrique indienne. Les minorités d'origine indo-pakistanaise en Afrique orientale*, Paris, IFRA/Karthala, 81–107.

50 L'article de N. Um, « Chairs, Writing Tables, and Chests: Indian Ocean Furniture and the Postures of Commercial Documentation in Coastal Yemen, 1700–40 », 2015, est consacré au mobilier lié aux écritures de la Cie néerlandaise des Indes orientales : au-delà de l'affirmation d'un standing, il y est question des nuées de scribes et *kātib*s à l'ouvrage, dont des Baniyans, Hindis et Jaïns du Gujerat, accompagnant l'activité de la Cie, des marchands privés et de l'administration du port, à al-Muḫā (Moka, Yémen) à une période antérieure.

51 R. K. P. Pankhurst, « The Trade of Northern Ethiopia in the Nineteenth and early Twentieth Centuries », janvier 1964, 142, d'après Plowden.

52 R. K. P. Pankhurst, « The Trade of Northern Ethiopia in the Nineteenth and early Twentieth Centuries », janvier 1964, 84 ; R. K. P. Pankhurst, « The Trade of Southern and Western Ethiopia and the Indian Ocean Ports in the Nineteenth and Early Twentieth Centuries », juillet 1965, 66. Voir la note 38 pour le Kenya.

53 R. K. P. Pankhurst, « The Trade of Southern and Western Ethiopia and the Indian Ocean Ports in the Nineteenth and Early Twentieth Centuries », juillet 1965, 65–66.

Adam, Michel. « Portraits et fragments d'histoires de vie au Kenya », dans : M. Adam (éd.), *L'Afrique indienne. Les minorités d'origine indo-pakistanaise en Afrique orientale*, Paris, IFRA/Karthala, 423–445.

Adjemian, Boris. 2013. *La fanfare du Négus : les Arméniens en Éthiopie (XIXe–XXe siècles)*, Paris, EHESS, « En Temps Et Lieux ».

Ahroni, Reuben. 1994. *The Jews of the British Crown Colony of Aden. History, Culture, and Ethnic Relations*, Leyde/New York/Cologne, E. J. Brill, « Jewish Studies » 12.

Asani, Ali Sultaan. 2002. *Ecstasy and Enlightenment: The Ismaili Devotional Literature of South Asia*, Londres/New York, I. B. Tauris, en association avec The Institute of Ismaili Studies, Londres.

Asani, Ali Sultaan. Juillet-septembre 1987. « The Khojkī Script: A Legacy of Ismaili Islam in the Indo-Pakistan Subcontinent », *Journal of the American Oriental Society* 107/3, 439–449.

Bezabeh, Samson Abebe. 2016. *Subjects of Empires/Citizens of States: Yemenis in Djibouti and Ethiopia*, Le Caire/New York, American University in Cairo Press.

Bhalloo, Zahir, & Iqbal Akhtar. À paraître. « Les manuscrits du sud de la vallée de l'Indus en écriture khojkī sindhī : état des lieux et perspectives », *Asiatische Studien/Études Asiatiques* 71/3.

Garretson, Peter P. 2000. *A history of Addis Abäba from its foundation in 1886 to 1910*, Wiesbaden, Harrassowitz, coll. « Aethiopistische Forschungen » 49.

Gori, Alessandro, coll. Anne Regourd, Jeremy R. Brown, Steve Delamarter. 2014. *A Handlist of the Manuscripts in the Institute for Ethiopian Studies*. Vol. 2 : *The Arabic Material of the Ethiopian Islamic Tradition*, Eugene (OR), Pickwick Publication, « Ethiopian Manuscrits, Texts, and Studies Series » 20.

Haines, Captain Stafford B. 1845. « Memoirs on the South and East Coast of Arabia », *Journal of the Royal Geographic Society* xv, 104–160.

Harre, Dominique. 2015. « The Indian Firm G. M. Mohamedally & Co in Ethiopia (1886–1937) », *Annales d'Éthiopie* 30, 285–311.

Kour, Zaka Hannah. 2004. *History of Aden, 1839–1872*, Londres/New York, Routlege. [Parution originale, 1981].

Le Roux, Hughes. 1902. *Menelik et nous : Le Carrefour d'Aden*, Paris, Librairie Nilsson.

Markovits, Claude. 1999. « Indian Merchant Networks outside India in the Nineteenth and Twentieth Centuries: A Preliminary Survey », *Modern Asian Studies* 33/4, Cambridge, Cambridge University Press, 883–911.

Miran, Jonathan. 2009. *Red Sea Citizens. Cosmopolitan Society and Cultural Change in Massawa*, Bloomingthon, Indiana University Press.

Pankhurst, Richard K. P., Amba, Aliu & Adjacent Markets. Juillet 1964. « The Trade of Central Ethiopia in the Nineteenth and Early Twentieth Centuries », *Journal of Ethiopian Studies* 2/2, 41–91.

Pankhurst, Richard K. P. Janvier 1964. « The Trade of Northern Ethiopia in the Nineteenth and early Twentieth Centuries », *Journal of Ethiopian Studies* 2/1, 49–159.

Pankhurst, Richard K. P. Janvier 1965. « The Trade of the Gulf of Aden Ports of Africa in the Nineteeth and Early Twentieth Centuries », *Journal of Ethiopian Studies* 3/1, 36–81.

Pankhurst, Richard K. P. Juillet 1965. « The Trade of Southern and Western Ethiopia and the Indian Ocean Ports in the Nineteenth and Early Twentieth Centuries », *Journal of Ethiopian Studies* 3/2, 37–74.

Pankhurst, Richard K. P. Janvier 1967. « Menilek and the Utilisation of Foreign Skills in Ethiopia », *Journal of Ethiopian Studies* 5/1, 29–86.

Pankhurst, Richard K. P. 1974. « Indian Trade with Ethiopia, the Gulf of Aden and the Horn of Africa in the Nineteenth and Early Twentieth Centuries », *Cahiers d'études africaines* 14/55, 453–497.

Pankhurst, Richard K. P. 1995. « The role of Indian craftsmen in late nineteenth and early twentieth-century Ethiopian palace, church and other building », *Journal of the Royal Asiatic Society* 5/1, 11–20.

Pankhurst, Richard K. P. 2002. « Across the Red Sea and Gulf of Aden: Ethiopia's Historic Ties with Yaman », *Africa* 57/3, 393–419.

Phillips Paper Trade Directory of the World. 1906/1907–1946/1947. London, S. C. Phillips & Co.

Ramaseshan, Sita. 1989. « The History of Paper in India up to 1948 », *Indian Journal of History of Science* 24/2, 103–121.

Regourd, Anne. 2014. « Introduction to the Codicology: Papers, Ruling, and Bindings », dans : Alessandro Gori & coll., voir à cette entrée, xlix–xcv.

Regourd, Anne. 2015. « Papiers filigranés de manuscrits de Zabīd, premier tiers du XVIIIe–milieu du XXe siècle : papiers importés et 'locaux' », Chap. 8 dans : David Hollenberg, Christoph Rauch & Sabine Schmidtke, *The Yemeni Manuscript Tradition*, Leyde/Boston, E. J. Brill, 227–251.

Trevaskis, Sir Kennedy. 1968. *Shades of Amber: A South Arabian Episode*, Londres.

Um, Nancy. 2015. « Chairs, Writing Tables, and Chests: Indian Ocean Furniture and the Postures of Commercial Documentation in Coastal Yemen, 1700–40 », *Art History* 38/4, 718–731.

Wagner, Ewald. 2003. *Harar. Annotierte Bibliographie zum Schrifttum über die Stadt und den Islam in Südostäthiopien*, Wiesbaden, Harrassowitz.

Wairire, Gidraph G. 2010. « Commerçants et industriels d'origine indo-pakistanaise au Kenya. Regards sociologiques », dans : M. Adam (éd.), *L'Afrique indienne. Les minorités d'origine indo-pakistanaise en Afrique orientale*, Paris, IFRA/Karthala, 277–298.

Wylde, Augustus. 1901. *Modern Abyssinia*, London, Methuen & Co.

Zervos, Adrien. 1936. *L'empire d'Éthiopie*, Alexandrie.

CHAPTER 6

Note sur les papiers à timbre sec (*dry seal*) en russe ou en arménien (second tiers du xixe siècle)

Francis Richard

Abstract

In this note attention is paid to papers bearing dry seals with Cyrillic inscriptions which were exported to Persia from the Russian Empire. From a brief look at such use in the Ottoman world, it can be shown that this paper was used in other places. Later this type of seal seems to have been used for authenticating documents.

1 Introduction

La production papetière issue de moulins installés en Russie, notamment dans les régions d'Uglich, Yaroslavl', Viatka, Moscou ou St. Pétersbourg, a donné lieu à des études détaillées et à la publication de répertoires de filigranes[1]. Ces papiers, vergés, souvent colorés en bleu, ont été d'un emploi fort fréquent, surtout en Perse depuis la fin du xviiie s. Il semble qu'ils aient été importés par rames *dasteh*, grâce aux négociants de la Caspienne. On les trouve employés jusque dans les années 1850 pour la copie des actes et documents, des livres manuscrits ou pour la confection des premiers ouvrages imprimés persans.

Le propos de cette note est d'aborder une nouvelle génération de papiers, apparus au second tiers du xixe s., de même provenance et destination.

2 Papiers dotés de timbres secs à caractères cyrilliques

Après le milieu du xixe s., on utilise très souvent, en Perse surtout, mais aussi dans l'Empire ottoman, des papiers de production industrielle, non-vergés, importés de Russie qui présentent la particularité d'être marqués au moyen d'un timbre sec imprimé sur chaque feuille. Les timbres évoquent par leur

1 Z. V. Uchastkina, *A History of Russian Hand Paper-Mills and their Watermarks*, 1963; K. Y. Tromonin, *Tromonin's Watermark Album*, 1965.

 | DOI 10.1163/9789004360877_008

aspect les timbres secs marqués « Bath » que l'on trouve sur des papiers anglais non-vergés de la même époque.

Parmi les papiers marqués d'un timbre sec à inscription en caractères cyrilliques que nous avons pu relever sur des manuscrits certains portent une date, d'autres un nom de fabrique ou des initiales. Ils sont souvent fort difficiles à lire car la marque a été effacée. Les papiers sont encore parfois colorés en bleu ou en vert, ce qui correspond au goût persistant des utilisateurs moyen-orientaux pour les papiers colorés.

A. Dans les collections de la Bibliothèque nationale de France

Parmi les manuscrits persans, nous avons pu relever les exemples suivants :

- dans le Supplément persan 996, collection de chants populaires du Khorassan collectés et transcrits par l'ancien Consul Alexandre Chodzko (1804–1891)[2], petit recueil de pièces copiées entre 1835 et 1852, se trouve un timbre sec russe de forme ovale.
- dans le Supplément persan 1131, un autre manuscrit venant de Chodzko, dont la partie la plus ancienne, cadeau de l'Imâm Qulî Mîrzâ[3], date de 1833, mais qui a été complété plus tard, on rencontre sur quelques feuillets un timbre sec marqué *Asvevskoi Fabriki A n° 5*.
- dans le Supplément persan 1061 à 1065, copie en plusieurs volumes réalisée à Téhéran entre 1853 et 1855 des *Rumûz-i Hamza* provenant de la bibliothèque d'Arthur de Gobineau[4], on peut voir aussi des timbres secs russes.
- dans le Supplément persan 1047, qui vient aussi de la bibliothèque d'A. de Gobineau, on trouve deux textes copiés en 1857–1858 sur un papier bleu portant différents timbres secs à marque russe[5].

---

2 E. Blochet, *Catalogue des manuscrits persans de la Bibliothèque nationale*, 1905–1934, vol. 3, n° 2015 ; voir aussi F. Richard, *Catalogue des manuscrits persans, Bibliothèque nationale de France, Département des manuscrits*. Vol. 2 : *Supplément persan. Deuxième partie 525–1000*, 2013, 1312–1313.

3 E. Blochet, *Catalogue des manuscrits persans de la Bibliothèque nationale*, 1905–1934, I, n° 685.

4 E. Blochet, *Catalogue des manuscrits persans de la Bibliothèque nationale*, 1905–1934, IV, n° 2106–2110.

5 E. Blochet, *Catalogue des manuscrits persans de la Bibliothèque nationale*, 1905–1934, IV, n° 2184.

- dans le Supplément persan 1055, une copie du poème *Yûsuf va Zulaykhâ* de Shamsî ʿErâqî[6] datée de 1858 et provenant encore de la bibliothèque d'A. de Gobineau, on rencontre des timbres secs russes à marque *A.C.*
- datant de 1852, le manuscrit Supplément persan 1124 est une copie du Dîvân de Qa'ânî[7] réalisée sur un papier avec différents timbres secs dont certains marqués *Nevskoï Fab.*
- le manuscrit Supplément persan 1444, venant de la collection Charles Schefer[8], une copie de l'histoire de Hérat (*Rawzat al-jannât*) faite en 1878 par le copiste Mahmûd Tabâtabâ'î Isfahânî, a été réalisé sur des feuilles de papier où l'on trouve des timbres à marque *Fabrika Sergyeva N.6*, ainsi qu'à marque ... *Kompanii Uglichskoï fabriki.*
- le manuscrit Supplément persan 1210, une Anthologie qui vient de James Darmesteter[9] est rédigée sur un papier où l'on trouve différents timbres secs. Il est du milieu du xixe s. mais n'est pas daté.
- un autre volume, le recueil Supplément persan 2121, contenant des pièces manuscrites et lithographiées, dont une partie est datée de 1868, contient des papiers à timbres secs russes[10].

B. Dans les collections de la Bibliothèque universitaire des langues et civilisations (BULAC, Paris)

On trouve différents timbres secs à inscription cyrillique – dont une marque *Fabriki Aristarhova 3* – par exemple, dans le manuscrit Pers. 133, « Abécédaire ou manuel d'écriture » composé par Qāpūtān Mīrzā Fatīh-ʿAlī b. Mīrzā Muh.-Taqī Āẖūnzāda (1812–1878), dit Mirza Akhundov[11], au mois de ṣafar 1274/septembre-octobre 1857 [Ill. 6.1].

On en trouve un autre exemple avec une des pièces (datable des alentours de 1855 ?) du recueil Pers. 138 de la BULAC [Ill. 6.2], ou avec la lettre n° 5 du manuscrit Pers. 148, un billet d'invitation rédigé à Tabriz, en date du 20 raǧab 1286/26 octobre 1869.

---

6 E. Blochet, *Catalogue des manuscrits persans de la Bibliothèque nationale*, 1905–1934, III, n° 1178.

7 E. Blochet, *Catalogue des manuscrits persans de la Bibliothèque nationale*, 1905–1934, III, n° 1950.

8 E. Blochet, *Catalogue des manuscrits persans de la Bibliothèque nationale*, 1905–1934, I, n° 509.

9 E. Blochet, *Catalogue des manuscrits persans de la Bibliothèque nationale*, 1905–1934, III, n° 2012.

10 Notice consultable en salle de lecture du Dpt. des Manuscrits.

11 Sur ces collections, voir http://bulac.hypotheses.org/333.

ILLUSTRATION 6.1 *Ms. Pers. 133*, BULAC

ILLUSTRATION 6.2 *Ms. Pers. 138*, BULAC

Avec ces quelques exemples, on constate que ces timbres secs marquant des papiers importés de Russie en Perse se rencontrent surtout entre 1850 et 1880.

3 Perpétuation de l'usage des timbres secs en Iran

L'usage de ces timbres secs s'est perpétué en Iran, notamment pour marquer le papier destiné à des emplois officiels. Ainsi on trouve dans le manuscrit Supplément persan 1538 – un inventaire de la bibliothèque du Vizir I'tizâd al-saltaneh, datant de 1293/1876, un timbre sec iranien à motif de soleil et de lion, mais sans inscription.

4 Timbres secs à caractères arméniens

On rencontre aussi un autre exemple intéressant de timbre sec en écriture non-européenne sur le ms. Turc 135 de la BULAC. Il s'agit d'une copie du milieu du XIXe s. des Dix *Maqâle* ou « articles » sur les réformes de l'Empire ottoman de Hamd-ullāh Afandī Ğazā'irlī : de forme ovale il porte, en caractères arméniens, la marque « *H. Tschētsahirlean* [i. e. Ğazā'irlī], [*Qostantino*]*Polis* » [Ill. 6.3], qui est celle de l'auteur.

ILLUSTRATION 6.3 *Filigrane, ms. Turc 135, BULAC*

5 Conclusion

C'est entre 1850 et 1880 que des papiers à timbre sec en caractères cyrilliques apparaissent dans des manuscrits en persan. Loin d'être un phénomène isolé, ils sont nombreux, laissant supposer l'existence de différents fabricants exportant régulièrement des rames de papier de Russie en Perse. Ces papiers sont utilisés dans tout le pays et supplantent de plus en plus les papiers locaux pour les besoins des copistes de manuscrits comme des imprimeurs. L'usage de timbres secs a perduré ensuite en Iran pour les documents officiels. Au milieu du xixe s., l'Empire ottoman a aussi importé ce type de papier de Russie, mais des manuscrits attestent l'existence d'autres papiers à timbre sec, portant inscriptions en caractères arméniens, où la marque pourrait avoir un autre usage que celui d'indiquer le lieu de fabrication.

Bibliographie

Blochet, Edgar. 1905–1934. *Catalogue des manuscrits persans de la Bibliothèque nationale*, Paris, Leroux, Réunion des Bibliothèques nationales, 4 vols.

Richard, Francis. 2013. *Catalogue des manuscrits persans, Bibliothèque nationale de France, Département des manuscrits*. Vol. 2 : *Supplément persan. Première partie, 1–524 ; Deuxième partie, 525–1000*, Rome, Istituto per l'Oriente C. A. Nallino & UMR 7528 « Mondes iranien et indien », « Series Catalogorum » V/1–2, 2 vols.

Tromonin, Kornilii Yakovlevich. 1965. *Tromonin's Watermark Album. A Facsimile of the Moscow 1844 edition with additional materials by S.A. Klepikov*, Hilversum, The Papers Publication Society, « Monumenta chartae papyricae historium illustrantia » XI.

Uchastkina, Zoâ Vasil'evna. 1963. *A History of Russian Hand Paper-Mills and their Watermarks, edited and adapted for publication in English by J.S.G. Simmons*, Hilversum, The Papers Publication Society, « Monumenta chartae papyricae historium illustrantia » IX.

CHAPTER 7

Un exemple rare de contremarque du viiie/xive siècle en langue et caractères arabes*

Alice Shafi-Leblanc

Abstract

According to the researches that have been led on the subject until today, watermarks and countermarks only appear in the 10th/16th century through imported European papers. However, as I was studying Moẓaffarid manuscripts illuminations, I happened to notice the presence of a countermark in Arabic language and script in a 8th/14th century Qur'an made in Šīrāz. It is read *al-ǧalālī* and could refer to a paper size or a workshop. Even if we still can't explain the existence of such a mark, as it is unique in this region and at that time, the fact remains that the lack of systematic paper analyses—which are often materially impossible—prevents us from knowing the extent of the phenomenon. Whether it be an isolated production or not, this countermark could show an early appropriation of European watermarked paper making techniques in the Persian world.

1 Introduction

La contremarque en langue et caractères arabes dont il est question dans cet article a été identifiée dans des manuscrits produits sous les Moẓaffarides, une dynastie qui régna principalement sur le Fārs entre 754/1353 et 793/1393–1394, lors de recherches menées sur l'étude de leurs décors enluminés[1]. Il s'agit d'un cas exceptionnel dans le monde persan de cette époque.

Nous débuterons par l'étude des manuscrits d'époque moẓaffaride concernés et de leur papier, puis nous situerons ce papier au regard de la production dans le même contexte pour, enfin, discuter sa provenance.

* Cet article n'aurait pu voir le jour sans la précieuse contribution d'Anne Regourd, qui m'a non seulement donné la formidable opportunité de le publier, mais m'a permis de le discuter plus avant en regard du domaine de recherche dans lequel il s'inscrit. Qu'elle soit remerciée ici de ses avis, relectures et corrections.

1 Ces recherches ont été conduites en deuxième année de Master à l'Université de Paris IV-Sorbonne, sous la direction de Jean-Pierre Van-Staëvel et Eloïse Brac de la Perrière.

 | DOI 10.1163/9789004360877_009

2 Une contremarque en alphabet arabe

Une contremarque en alphabet arabe a été relevée dans un Coran en trente sections (*ǧuz*ʾs), dont les sections 21 et 30 sont conservées à Londres, respectivement au Victoria & Albert Museum (dorénavant V & A ; ms. 361–1885) et dans la Collection Khalili (ms. QUR 159). Ces *ǧuz*ʾs sont datables de l'époque moẓaffaride. En effet, leurs reliures et leurs décors sont semblables à ceux des volumes du Coran commandé par la princesse inǧue Fārs Malik Ḫātūn (vivante ca. 749/1348) (Collection Khalili, ms. QUR 181/Širāz, Pars Museum, ms. 417), enluminés et reliés sous le vizir moẓaffaride Turān Šāh à Širāz vers 777/1375, suivant la date de la *waqfiyya* établie à son nom[2]. Les reliures de tous ces volumes, en cuir brun lisse, arborent un médaillon central en toupie dessiné à l'or et flanqué de pendentifs en fleurons. Au centre, un motif estampé évoquant un svastika est pris dans un cercle doré, point de départ d'un décor géométrique étoilé. Les espaces ménagés par les filets dorés sont remplis par un décor estampé de tresses et d'étoiles aux angles arrondis. Le pourtour des plats est parcouru par une tresse estampée cernée de deux liserés dorés. Les écoinçons sont décorés d'un motif géométrique doré composé d'entrelacs. Le médaillon central de la reliure du Coran de Fārs Malik Ḫātūn est quelque peu différent, mais il contient des motifs relevant d'un répertoire voisin. L'organisation du décor – page de frontispice avec bandeaux de titres bordant la zone de texte et empiècement marginal triangulaire –, ainsi que les motifs et la palette de couleur – cartouche central flanqué de cantons triangulaires, ornement de rameaux de feuilles dorées sur fond bleu lapis – permettent d'avancer que le volume du V & A et son pendant de la Collection Khalili relèvent du même type de production, c'est-à-dire qu'ils furent probablement enluminés à Širāz vers 777/1375. La date de leur copie est incertaine : sans conteste antérieure aux enluminures, elle n'est pas nécessairement aussi ancienne que celle du ms. QUR 181/ms. 417, exécutée après 736/1336 et plus vraisemblablement entre 743–744/1343 et 758/1357[3]. Il est donc possible de situer la production de ce Coran quelque part entre 743–44/1343 et 777/1375.

2 D. James, *The Master Scribes, Qur'āns of the 10th to the 14th Centuries AD*, 1992, 123 et 130.

3 D. James, *The Master Scribes*, 1992, 123. James explique que la première *waqfiyya* du manuscrit fait mention du « défunt » MaḥmūdŠhāh, père de Fārs Malik Ḫātūn, qui régna sur Širāz et mourut en 736/1336. La *waqfiyya*, placée sur la page initiale du manuscrit et accompagnée d'enluminures, est aussi ancienne que l'objet. La princesse inǧue qui en a fait la commande avait également précisé que le manuscrit resterait chez elle jusqu'à sa mort, puis serait placé dans sa tombe. La copie est donc nécessairement postérieure à 736/1336. Par ailleurs, outre les dates de règne d'Abū Esḥāq qui marquent une période de relative stabilité propice au mécénat (entre 736/1336 et 743/1343, des guerres opposent les fils de MaḥmūdŠhāh à des dynasties rivales, tandis qu'ils se livrent aussi des guerres fratricides, voir http://www

Or l'examen du papier de l'exemplaire du V & A, le *ǧuz'* 21, a révélé une inscription en alphabet arabe (ou arabo-persan) qui se lit *al-ǧalālī*[4]. Celle-ci se trouve dans le coin de la feuille, en bas de la marge de gouttière, à onze reprises sur un total de vingt-trois bifeuillets – soit presque la moitié – ce qui indique que les bifeuillets pourraient correspondre à la moitié des feuilles originelles. Les dimensions des feuillets sont de 36,5 × 28,2 cm. Le volume est formé de quaternions, avec un premier cahier lacunaire, comptant deux puis quatre feuillets[5]. Le volume devait donc compter 24 bifeuillets à l'origine[6]. Parallèle aux vergeures, l'inscription est visible uniquement par transparence : aucune marque n'apparaît à la surface du papier et il n'y a aucune anomalie dans la trame qui pourrait laisser croire à un ajout. Sur les folios 20b et 24b, une vignette marginale dissimule en partie la contremarque, montrant ainsi que le décor est bien postérieur à celle-ci. Quant à son pendant de la Collection Khalili (ms. QUR 159), la contremarque n'apparaît en revanche que cinq fois sur un total de vingt-six bifeuillets, en haut de la marge de gouttière[7]. La nature et l'emplacement de cette inscription, qui ne se retrouve jamais deux fois sur le même bifeuillet, rendent compte du choix du terme de contremarque pour la désigner[8]. Le volume de la Collection Khalili ne compte que cinq occurrences de cette contremarque, mais un autre papier a pu être utilisé en complément pour constituer le manuscrit.[9] On peut aussi supposer que, comme il s'agit d'un Coran en trente volumes, le papier préparé et découpé en amont ait été

.iranicaonline.org/articles/inju-dynasty), la dernière mention de Fārs Malik Ḫātūn vivante dans les textes pourrait être celle des *Voyages* d'Ibn Battuta, en 749/1348, cf. *The Travels of Ibn Battuta*, by H.A.R. Gibb, 1962, vol.1, 305. La date de début de copie du Coran de Fārs Malik Ḫātūn pourrait donc être située entre 743–744/1343 et 749/1348, puis la copie ainsi que la première phase d'enluminures ont dû prendre fin avec la chute de la dynastie inǧue en 758/1357.

4 Je remercie Tim Stanley, conservateur en chef du département Moyen-Orient au V & A, de l'aide qu'il m'a fournie pour déchiffrer l'inscription.

5 Le *ǧuz'* 21 commence au milieu du v. 46 de la sourate 29, al-ʿAnkabūt/l'araignée. Les deux feuillets du premier cahier, manquants, devaient abriter le début de la sourate. On peut d'ailleurs supposer que les deux feuillets manquants comprenaient un autre feuillet de frontispice et un *šamse* ou toute autre forme de feuillet de garde.

6 Et peut-être bien plus, car le *ǧuz'* s'achève au milieu du v. 28 de la sourate 33, al-Aḥzāb/les factions, qui compte normalement soixante-treize versets.

7 Tim Stanley et Nahla Nasser, conservatrice de la Collection Khalili, ont trouvé la trace de la contremarque dans le QUR 159 et m'en ont aimablement transmis les clichés, ce dont je leur suis très reconnaissante. La contremarque apparaît sur les ff. 6a, 13a, 38a, 43a et 52a.

8 Voir *infra*, section VI.

9 F. Déroche *et alii, Manuel de codicologie des manuscrits en écriture arabe,* 2000, 56–57.

réparti indifféremment entre les *ǧuz*'s, rendant tout aussi inégale la distribution des contremarques. En revanche, la présence d'une contremarque incite à rechercher un filigrane. Après vérification, il n'y en a aucun dans les deux *ǧuz*'s de ce Coran. Toutefois, cette marque a pu être éliminée lors de la découpe du papier ou s'être trouvé masquée par la reliure, si elle était placée au centre de la

ILLUSTRATION 7.1 *V & A Museum (ms. 361–1885), f. 26b*

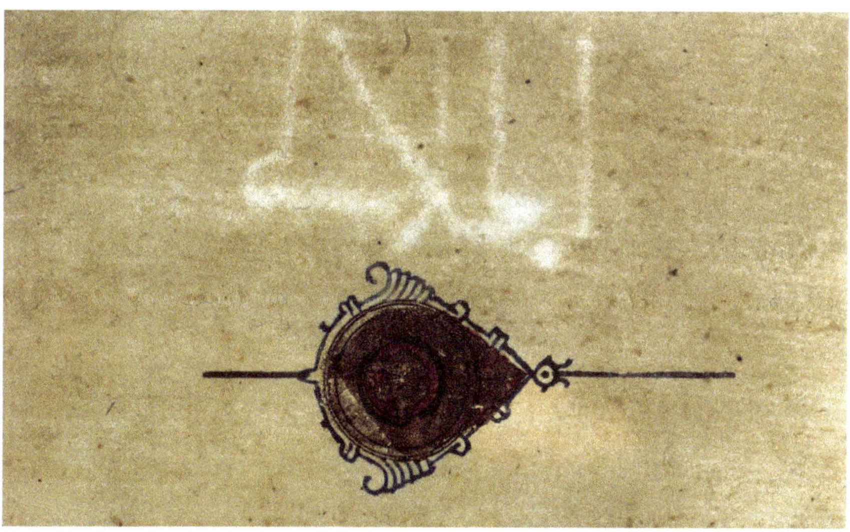

ILLUSTRATION 7.2 *Collection Khalili (QUR 159), f. 13a*

feuille d'origine et de petite dimension[10]. Sans pour autant affirmer qu'il s'agit bien là d'une contremarque, il est possible de recevoir cette hypothèse, étant donné la position et la nature de la marque *al-ǧalālī*.

3 Caractéristiques physiques de la contremarque

Afin de comparer les contremarques, des mesures ont été relevées sur le manuscrit du V & A. Ont été prises en compte la hauteur et la largeur du champ réservé à l'inscription *al-ǧalālī*, ainsi que la longueur des hampes des lettres et de leurs parties inférieures – dont l'espacement entre la base légèrement coudée de l'*alif* et le point du *ǧīm* –, la longueur de la section inférieure du *ǧīm*, la section inférieure gauche du *lām-alif*, la longueur des sections supérieure et inférieure du *ī* d'*al-ǧalālī*, ainsi que la hauteur du retour incurvé.

Le cliché ci-dessous reprend ces éléments. Une lettre de référence est donnée à chaque mesure, afin de clarifier les résultats reportés dans le tableau qui fait suite :

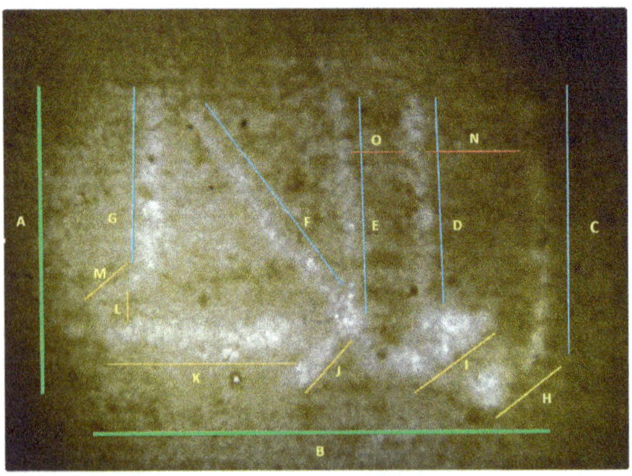

ILLUSTRATION 7.3 *Ms. 361–1885, V & A. Mesures relevées sur la contremarque, référencées de A à O*

10 Au sujet de la place du filigrane dans la feuille en fonction du format et des vergeures, voir J. Irigoin, « La datation par les filigranes du papier », 1980, 13.

Tableau récapitulatif des mesures effectuées sur les contremarques (en millimètre et degré)

	A	B	C	D	E	F	G	H	I	J	K	L	M	N	O	Angle E–F[a]
F. 2b	28	42	20	17	17	20	16	7	5	7	18	5	5	10	5	40°
F. 3b	29	42	23	20	18	20	21	7	9	7	16	3	4	9	5	40°
F. 7b	29	42	23	18	19	24	12	6	13	7	18	5	6	9	5	40°
F. 20b	27	42	X	X	20	24	18	X	X	8	18	2	5	9	4	40°
F. 22b	28	43	23	17	19	22	17	6	9	8	18	1	5	9	5	40°
F. 24b	28	42	23	20	X	X	18	5	X	X	X	4	5	9	4	40
F. 26b	29	41	23	20	18	20	18	5	11	7	18	4	5	9	4	40°
F. 35b	29	42	24	19	19	25	18	5	6	7	19	2	4	9	4	40°
F. 38b	28	42	23	20	20	22	18	6	10	5	18	3	4	9	5	40°
F. 44b	29	42	22	18	20	22	18	6	12	6	18	2	5	9	5	40°
F. 46b	29	42	24	18	20	22	18	5	12	6	18	2	5	9	5	40°
Moyennes (au mm supérieur)	29	42	23	19	19	22	18	6	10	7	18	3	5	9	5	40°
Valeur médiane	28	42	22	18,5	18,5	22,5	16,5	6	9	6,5	17,5	3	5	9,5	4,5	
Variation maximale en mm[b]	1	1	2	1,5	1,5	2,5	4,5	1	4	1,5	1,5	2	1	0,5	0,5	
Variation maximale en % (arrondi à la 1ère décimale)[c]	3,6	2,4	9,1	8,1	8,1	11,1	27,3	16,6	44,4	23,1	8,6	66,7	20	5,3	11,1	

a Il peut y avoir une marge d'erreur d'environ deux degrés, compte tenu des conditions d'observation, de la lisibilité de la contremarque elle-même et de l'épaisseur de son tracé. Les angles font donc tous, à deux degrés près, quarante degrés.

b Il s'agit de la variation maximale entre la valeur médiane (et non pas la moyenne des valeurs) et la valeur minimale ou maximale, ce qui exprime une marge d'erreur, de + ou – 1 mm (par ex. la mesure A = ± 28 mm). Le calcul est communément appelé « tolérance dimensionnelle ». Pour obtenir la valeur médiane, il faut additionner la valeur maximale et la valeur minimale, puis diviser ce total par deux (valeur médiane = (valeur max + valeur min) ÷ 2). Ensuite, pour calculer la variation maximale, il faut soustraire la valeur minimale de la valeur maximale et diviser ce résultat par deux (variation max = (valeur max-valeur min) ÷ 2).

c Il s'agit d'exprimer la variation maximale en pourcentage (par exemple A = 28 mm = + ou – 3,6 %). On obtient ce pourcentage en multipliant la variation maximale par cent, puis en divisant ce résultat par la valeur médiane (pourcentage = (variation max × 100) ÷ valeur médiane).

N.B. : les champs remplis par X correspondent à des données non observables en raison de l'emplacement du décor.

Malgré la relative homogénéité de la taille du champ, des lettres et de leur espacement – et notamment l'angle E-F qui mesure à chaque fois 40 degrés, il convient de noter qu'en plus d'un taux de variation important sur certaines mesures, la forme même des lettres varie d'un feuillet à l'autre [ill. 7.4 à 7.8]. Il est par conséquent difficile d'envisager que la contremarque ait été fabriquée à l'aide d'une pièce unique et solide, tel un tampon, qui aurait laissé, lui, une marque immuable. En revanche, le pressage de chaque feuille d'un papier produit à partir d'un moule à fils métalliques, tels ceux employés dans les villes italiennes, aurait pu entraîner que les marques ainsi réalisées aient bougé. Mais n'en résulterait-il pas plutôt un « floutage » de l'inscription qu'une modification des formes ? Or les lettres sont ici clairement délimitées, à quelques exceptions près, par exemple le trait inférieur du *ī* d'*al-ǧalālī* sur les folios 26b et 38b [ill. 7.4 et 7.6]. Reste que les variations pourraient s'expliquer par l'utilisation de plusieurs moules, et, dans la mesure où elles sont nombreuses, être l'indication que le papetier avait de nombreux moules et produisait beaucoup. Dans cette hypothèse, il ne s'agit pas d'une petite entreprise.

4 État des lieux : un cas unique

Dans l'état de nos connaissances, les premiers papiers filigranés ont été élaborés en Italie. Le plus ancien date de 1282[11]. Puis les contremarques, généralement formées des initiales ou du nom du fabricant de papier, parfois accompagnées de la date et du lieu de production, ont été ajoutées au xe/xvie s.[12], soit dans le coin supérieur gauche, soit au milieu de la moitié gauche de la feuille[13], ou bien sous le filigrane.

Les papiers italiens filigranés ont pénétré le Maghreb dès le milieu du viiie/xive s.[14]. On retrouve sur ces destinations des papiers dont le répertoire iconographique n'est pas spécifique : les symboles présentant un rapport avec le monde musulman – tels que les croissants de lune – n'ont définitivement envahi le marché qu'au xie/xviie s.[15]. Quant aux contremarques en caractères arabes ou en alphabet turc-ottoman connues à ce jour, on ne les trouve qu'à

11 Ch.-M. Briquet, *Les filigranes. Dictionnaire historique des marques du papier dès leur apparition vers 1282 jusqu'en 1600*, 1907, vol. 1, 10.

12 F. Déroche *et alii, Manuel de codicologie,* 2000, 66.

13 A. Gacek, *Arabic Manuscripts: A Vademecum for Readers*, 2009, « Watermarks », 290–292.

14 F. Déroche *et alii, Manuel de codicologie,* 2000, 64.

15 A. Gacek, *Arabic Manuscripts,* 2009, « Watermarks », 291 ; A.Velkov & S. Andreev, *Filigranes dans les documents ottomans.* Vol. 1 : *Trois croissants*, 1983. Un papier *Tre lune* a

ILLUSTRATION 7.4
F. 7b

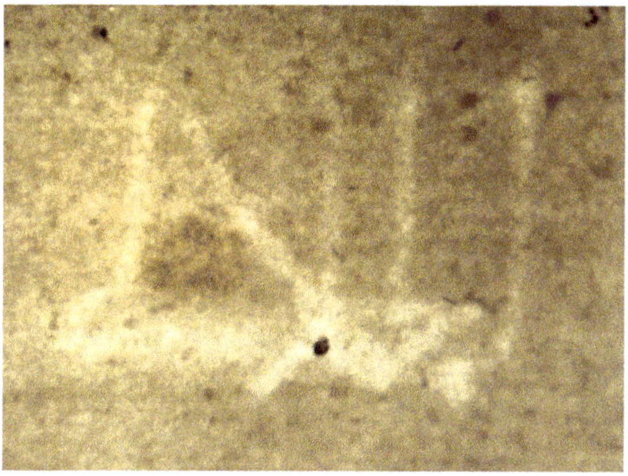

ILLUSTRATION 7.5
F. 26b

ILLUSTRATION 7.6
F. 35b

ILLUSTRATION 7.7
F. 38b

ILLUSTRATION 7.8
F. 44b

partir du xiiie/xixe s., soit en langue *osmanlı* dans du papier ottoman fabriqué selon une technique italienne ou dans du papier vélin français, soit en caractères arabes et différentes langues sur des papiers à timbre sec et des papiers européens faits à la machine[16].

Dans le monde persan, aucun papier européen filigrané n'a été répertorié à ce jour avant le xe/xvie s. et ils sont encore peu nombreux jusqu'à la fin du xiie/xviiie s.[17]. Un des rares exemples antérieurs concerne un manuscrit de

---

été récemment identifié dans un manuscrit yéménite daté au colophon de 1045/1636, le ms. Glaser 178, Staatsbibliothek zu Berlin (communication d'Anne Regourd).

16 G. Humbert, « Le manuscrit arabe et ses papiers », 2002, 40, 44 ; A. Gacek, *Arabic Manuscripts,* 2009, « Impressed watermarks », 130–131 ; et dans ce volume les contributions de Michaelle Biddle (2), Evyn Kropf, Anne Regourd et Jan Just Witkam, n. d. l. e.

17 F. Déroche *et alii, Manuel de codicologie,* 2000, 65.

753/1352 conservé à la Bibliothèque nationale de France à Paris[18]. Son papier est génois[19] et les filigranes sont formés de figures géométriques. Rédigé en persan, il a cependant été copié à Sarāy-i Jadīd, en ancienne Crimée (Tsarev puis actuellement Kolobovka en Russie), alors territoire de la Horde d'Or[20]. Outre les contacts existant entre la Crimée et Gênes, qui possédait des comptoirs dans cette région, des échanges commerciaux entre les villes italiennes et Tabriz s'étaient également établis à la fin du viie/xiiie s. et au viiie/xive s.[21]. Ainsi, des papiers orientaux – tels que le papier élaboré à Tabriz sous les Ǧalayirides (1336–1432)[22] – étaient importés en Italie. En échange, il se peut donc que des papiers filigranés aient été introduits dans le monde persan. Cependant, jusqu'au ixe/xve s., l'Iran a essentiellement utilisé des papiers produits localement, exception faite de quelques probables importations de Samarqand et du Ḫatāy[23].

Quoiqu'il en soit, aucune contremarque en langue et caractères arabes antérieure au xiiie/xixe s. n'a été jusqu'ici répertoriée dans les catalogues et les bases de données spécialisées, y compris ceux consacrés aux manuscrits arabes, persans et ottomans[24].

5 Provenance du papier

Jean-Louis Estève, dont les recherches sur les caractéristiques matérielles des papiers fabriqués dans le monde arabo-islamique entre 133/751 et 750/1350 sont connues, a proposé de vérifier la constance du diamètre des vergeures pour

18 Supplément Persan 113. Voir F. Richard, *Catalogue des manuscrits persans, Bibliothèque nationale de France, Département des manuscrits*. Vol. 2 : *Supplément persan*, 2013, vol. 1, 168–171 ; http://archivesetmanuscrits.bnf.fr/ead.html?id=FRBNFEAD000090868&c=FRBNFEAD000090868_e0000016&qid=eas1423614258743.

19 F. Déroche *et alii*, *Manuel de codicologie*, 2000, 64.

20 C. P. Atwood, *Encyclopedia of Mongolia and the Mongol Empire*, 2004, 204.

21 Y. Porter, *Les Iraniens : Histoire d'un peuple*, 2006, 171 ; C. Delacroix-Besnier, *Les Dominicains et la chrétienté grecque aux XIVe et XVe siècles*, 1997, 20.

22 F. Richard, *Splendeurs persanes. Manuscrits du XIIe au XVIIe siècle*, 1997, 62.

23 Y. Porter, « Notes sur la fabrication du papier dans le monde iranien médiéval », 1999, 27.

24 Sans vouloir être exhaustif, on citera essentiellement ici, parmi les catalogues et bases de données en question, la collection Piccard en ligne http://www.piccard-online.de et http://www.ksbm.oeaw.ac.at/_scripts/php/PPO.php, le catalogue en ligne de Charles Briquet http://www.ksbm.oeaw.ac.at/_scripts/php/BR.php, Thomas L. Gravell Watermark Archive http://www.gravell.org, et la base du Dutch Institute for Art History de Florence http://www.wm-portal.net/niki/index.php.

vérifier si la forme est à base de végétaux ou de tiges métalliques, et déterminer si le papier est oriental ou si c'est une production occidentale[25]. Celle-ci pourrait alors avoir été soit destinée à être exportée à grande échelle vers l'Orient, soit plus spécifiquement à un atelier. Les vergeures sont verticales et relativement fines, puisque l'on en compte vingt pour vingt-deux millimètres. Observées à la feuille lumineuse, elles ne semblent pas rigoureusement régulières [ill. 7.9], mais faute de matériel adéquat, leur diamètre n'a pas pu être mesuré. Dans l'ensemble, l'observation à l'œil nu du papier ne remet pas *a priori* en cause son origine orientale et il est peu probable que des fils métalliques aient été employés pour réaliser la forme. Cela se vérifie également pour le ms QUR 159.

Si l'on voulait toutefois envisager l'hypothèse d'une fabrication italienne, cela supposerait que cette contremarque ait été élaborée et placée sur la forme en Italie, puis le papier produit exporté. Le scénario est historiquement plausible, car contremarques et filigranes ont été développés afin de rendre l'atelier ou le lieu de production des papiers reconnaissables et, ainsi, appliquer la taxe[26], mais aussi dans un contexte de concurrence croissante[27]. Cependant, aucun document, ni livre manuscrit portant une marque similaire n'ont été trouvés à ce jour pour corroborer cette hypothèse. Ce serait de fait un cas unique de papier occidental ancien à contremarque en langue et caractères arabes, i. e. qui ne serait ni fabriqué à la machine, ni porteur d'une contremarque en relief[28] ou imprimée.

Si au contraire le papier est bien oriental et réalisé à base de fibres végétales, comme le suggère l'observation à l'œil nu, on serait en présence d'un cas plus exceptionnel encore puisque, selon ce que nous savons, les contremarques n'avaient pas cours dans cette sphère géochronologique. Celles-ci sont même considérées comme incompatibles avec les techniques orientales de fabrication du papier, la forme souple empêchant l'insertion d'une marque rigide[29]. La codicologie des manuscrits orientaux étant une science relativement ré-

25 Communication personnelle de Jean-Louis Estève en mai 2011. La technique occidentale consiste en effet à insérer un fil de cuivre façonné selon la marque de l'atelier entre le moule et les fibres du papier ; G. Humbert, « Le manuscrit arabe », 2002, 40 ; D. Muzerelle, *Vocabulaire codicoloqique : répertoire méthodique des termes français relatifs aux manuscrits*, 1985 (version en ligne 2002–2003, http://vocabulaire.irht.cnrs.fr).

26 T. Walz, « The Paper Trade of Egypt and the Sudan in the Eighteenth and Nineteenth Centuries », 1985, 40.

27 G. Humbert, « Le manuscrit arabe et ses papiers », 2002, 40.

28 En anglais cette technique est appelée « imitation watermark », voir A. Gacek, *Arabic Manuscripts*, 2009, sous « Imitation watermarks », 130.

29 J. Irigoin, « La datation par les filigranes », 1980, 13 et 15.

cente, les nombreux manuscrits répertoriés n'ont pas nécessairement vu leur papier examiné méticuleusement. Plus encore, le présupposé conduisant les recherches, à savoir qu'il n'y a pas *a priori* de contremarque à trouver, a certainement fini par peser. En outre, rares sont les manuscrits qui possèdent encore leur reliure d'origine et qui n'ont pas été rognés. Même si l'usage de la contremarque a pu un tant soit peu se répandre – une volonté d'imiter les manuscrits occidentaux ? –, elles ont pu bel et bien disparaître au cours de l'une de ces opérations.

6 Interprétation de la contremarque *al-ǧalālī*

Les catalogues papier et informatique de contremarques ne répertorient rien de similaire, on l'a dit. *Al-ǧalālī* signifie « le corpulent, le grand » en arabe. Si cette inscription a bien une fonction de contremarque, compte tenu de son aspect, de son emplacement et de son occurrence par rapport au nombre de bifeuillets, elle devrait permettre d'identifier le nom de l'atelier de production. Le suffixe *–ī*, ou *nisba*, dans *al-ǧalālī*, pourrait se rapporter à un nom propre qui pourrait désigner par exemple le patron de l'atelier.

L'inscription pourrait également évoquer, si on retient la valeur d'adjectif de la *nisba*, le format du papier[30]. Les bifeuillets correspondant sans doute à la moitié de la feuille d'origine, cette dernière devait donc mesurer ca. 73 × 56,4 cm. Ajoutons à cela l'épaisseur relative de la feuille, il devient tout à fait possible que l'inscription désigne le papier lui-même. Celui-ci est par ailleurs d'une certaine qualité et ce grand format aurait alors été naturellement destiné à la réalisation de manuscrits prestigieux.

7 Conclusion

La datation du papier étudié ne peut être remise en cause étant donné le nombre d'éléments qui rattache ces volumes au QUR 181. Le décor, tant du point de vue de ses caractéristiques formelles que de sa qualité, peut être rapporté sans réserve au reste de la production moẓaffaride et il n'est pas envisageable que celui-ci soit une réplique tardive, « à la manière de ». La présence d'une contremarque de ce type est donc très surprenante et inédite pour le lieu

30 À ce sujet, G. Humbert cite al-Qalqašandī et note que dans le monde arabe – déjà à des périodes antérieures à notre manuscrit – les papiers pouvaient être désignés par leur format, voir « Le manuscrit arabe », 2002, 49 & 62.

ILLUSTRATION 7.9 *Ms. 361–1885, f. 20b*

et la période. Quant au QUR 181, pour lequel on aurait pu s'attendre à trouver une marque similaire, il n'en comporte aucune[31]. Il faudrait cependant vérifier son pendant de Šīrāz (ms. 417). Ceci soulève aussi le problème du réexamen de tout le corpus. En effet, si j'ai systématiquement procédé à l'analyse de plusieurs feuillets dans chacun des manuscrits avant de déterminer les caractéristiques du papier, chaque feuillet n'a pu être observé faute de temps d'abord, mais surtout parce que la recherche n'avait pas pour objectif les marques du papier. Plus radicalement, beaucoup de manuscrits ayant été rognés, il n'y a que peu d'espoir de trouver des contremarques placées à l'origine en marge.

Pourtant une telle étude – difficilement réalisable du fait de l'accessibilité, des conditions d'observation et du temps nécessaire – serait très profitable et même indispensable pour approfondir le sujet et tenter de trouver des réponses. Elle ne serait pas sans perspective, en cherchant notamment du côté de la littérature et du contexte historique. Il faudrait étendre l'étude à d'autres sphères du monde iranien[32], étudier de plus près la ville de Kermān, qui n'est un lieu important de culture qu'au viiie/xive s. et qui ne reprend ce statut qu'au xiie/xviiie s., et compléter ainsi l'enquête sur l'origine et le rayonnement de la peinture moẓaffaride. Il reste à chercher des réponses quant à la façon dont

31 Nahla Nasser a procédé par deux fois à l'examen du QUR 181 avant de me faire part de ses conclusions.

32 Francis Richard suggère de s'intéresser à la région du Khorassan – qui devint un centre florissant des arts du livre sous les Timourides grâce en partie aux artistes déplacés depuis Šīrāz – pour avoir peut-être un autre éclairage sur la production dans le Fārs au viiie/xive s.

les transformations, si particulières et pérennes qu'ont subies les arts du livre sous les Moẓaffarides, ont pu se produire, et dans quel environnement culturel.

Il ne reste plus à espérer que cet article ouvre la voie à de futures recherches.

Bibliographie

Atwood, Christopher Pratt. 2004. *Encyclopedia of Mongolia and the Mongol Empire*, New-York, Facts on File.

Briquet, Charles-Moïse. 1907. *Les filigranes. Dictionnaire historique des marques du papier dès leur apparition vers 1282 jusqu'en 1600*, Genève, A. Jullien, 4 vol.

Delacroix-Besnier, Claudine. 1997. *Les Dominicains et la chrétienté grecque aux XIVe et XVe siècles*, Rome, École française de Rome.

Déroche, François, avec la collaboration d'Annie Berthier, Marie-Geneviève Guesdon, Bernard Guineau, Francis Richard, Annie Vernay-Nouri, Jean Vezin, Muhammad Isa Waley. 2000. *Manuel de codicologie des manuscrits en écriture arabe*, Paris, Bibliothèque nationale de France, « Études et Recherches ».

Gacek, Adam. 2009. *Arabic Manuscripts: A Vademecum for Readers*, Leyde/Boston, E. J. Brill.

Humbert, Geneviève. Novembre 2002. « Le manuscrit arabe et ses papiers », *Revue des mondes musulmans et de la Méditerranée* 99–100 [mis en ligne le 12 mai 2009. URL : http://remmm.revues.org/1174].

Ibn Battuta. 1962. *The Travels of Ibn Battuta*, by H. A. R. Gibb, Londres, Hakluyt Society, 2 vols.

Irigoin, Jean. 1980. « La datation par les filigranes du papier », dans : Albert Gruys & Johan Peter Gumbert (éd.), *Les matériaux du livre manuscrit*, Leyde/Boston, E. J. Brill, « Codicologica, litterae textuales » V, 9–36.

James, David. 1992. *The Master Scribes, Qur'āns of the 10th to the 14th Centuries AD*, Londres, Azimuth Editions.

Muzerelle, Denis. 1985. *Vocabulaire codicologique : répertoire méthodique des termes français relatifs aux manuscrits*, Paris, CEMI [version en ligne 2002–2003, http://vocabulaire.irht.cnrs.fr].

Porter, Yves. 1999. « Notes sur la fabrication du papier dans le monde iranien médiéval », dans : Monique Zerdoun-Bat-Yehouda (éd.), *Le papier au Moyen-Âge : histoire et techniques*, Turnhout, Brepols, « Bibliologia. Elementa ad librorum Studia pertinentia » XIX, 19–30.

Porter, Yves. 2006. *Les Iraniens : Histoire d'un peuple*, Paris, Armand Colin.

Richard, Francis. 1997. *Splendeurs persanes. Manuscrits du XIIe au XVIIe siècle*, Paris, Bibliothèque nationale de France.

Richard, Francis. 2013. *Catalogue des manuscrits persans, Bibliothèque nationale de France, Département des manuscrits*. Vol. 2 : *Supplément persan. Première partie, 1–524 ; Deuxième partie, 525–1000*, Rome, Istituto per l'Oriente C. A. Nallino & UMR 7528 « Mondes iranien et indien », « Series catalogorum » V/1–2, 2 vols.
Velkov, Asparoukh & Andreev, Stephane. 1983. *Filigranes dans les documents ottomans*. Vol. 1 : *Trois croissants*, Sofia, Bibliothèque Nationale « Cyrille et Methode », Série Orientale [en Bulgare et Français].
Walz, Terence. 1985. « The Paper Trade of Egypt and the Sudan in the Eighteenth and Nineteenth Centuries », dans : Martin William Daly, *Modernization in the Sudan. Essays in Honor of Richard Hill*, New York, Lilian Barber Press, 39–41.

Publications en ligne

http://archivesetmanuscrits.bnf.fr/ead.html?id=FRBNFEAD000090868&c=FRBNFEAD000090868_e0000016&qid=eas1423614258743.
http://www.gravell.org.
http://www.iranicaonline.org/articles/inju-dynasty.
http://www.ksbm.oeaw.ac.at/_scripts/php/PPO.php.
http://www.ksbm.oeaw.ac.at/_scripts/php/BR.php.
http://www.piccard-online.de.
http://www.wm-portal.net/niki/index.php.

Crédits photographiques

Illustrations n° 1, 3, 4, 5, 6, 7, 8, 9 : clichés personnels, reproduits avec l'aimable autorisation du Victoria & Albert Museum, Londres.
Illustrations n° 2 et 10 : clichés de la Collection Khalili, Londres, reproduits avec l'aimable autorisation des responsables de la Collection.

Annexes : cartes

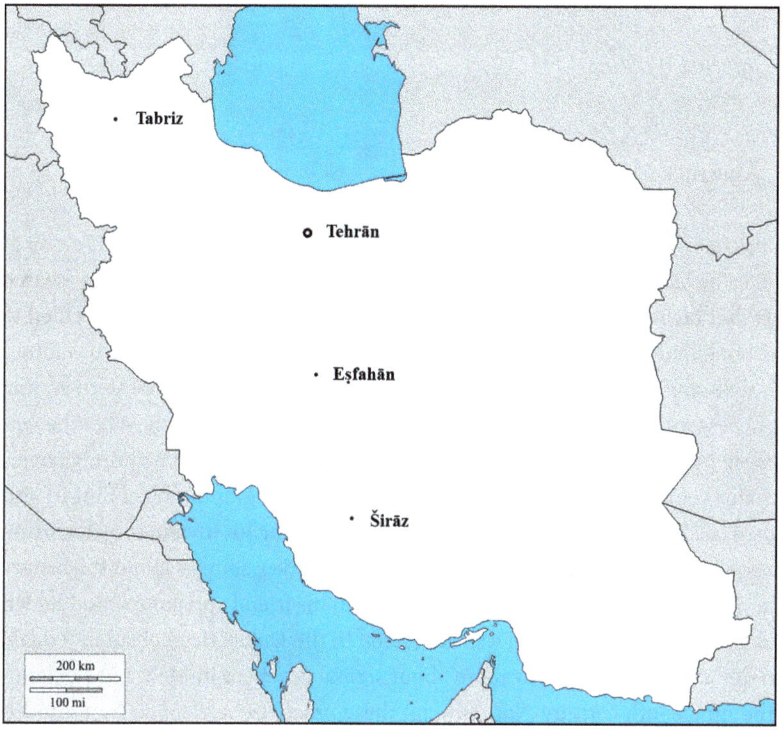

CARTE 7.1 *Principales villes d'Iran*
CARTE ALICE SHAFI-LEBLANC

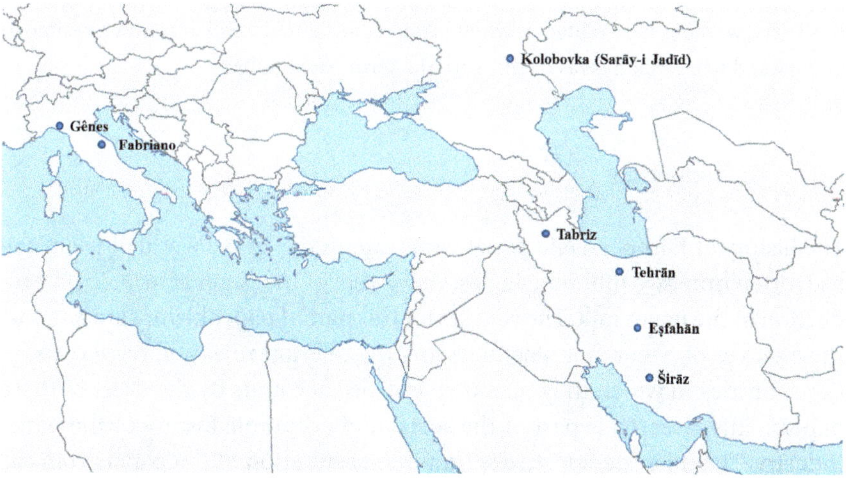

CARTE 7.2 *Principales villes citées*
CARTE ALICE SHAFI-LEBLANC

CHAPTER 8

Copy on demand. Abū Šubbāk in Mecca, 1303/1886

Jan Just Witkam

Abstract

When Christiaan Snouck Hurgronje (1857–1936) was exiled from Mecca and Arabia in August 1885, he had not even begun to think about what he would do with the research materials that he had been collecting. Back at his home base in Leiden he decided to write the book that would bring him instant fame, *Mekka*. It appeared in two volumes between 1888 and 1889. Snouck Hurgronje's abrupt departure from Western Arabia made it necessary for him to maintain lines of information on Mecca. When he was chased away by the Turkish authorities he had been on the point of acquiring copies of two major sources for a history of Mecca, the *Manā'iḥ al-Karam* by al-Sinǧārī and the *Ḫulāṣat al-Kalām* by Aḥmad b. Zaynī Daḥlān. Following his involuntary departure he managed to acquire handwritten copies of these two key sources about the history of Mecca. Through a recently discovered letter from his friend and companion Raden Aboe Bakar Djajadiningrat, which was addressed to the Dutch vice-consul in Jeddah, P. N. van der Chijs, we now know more about scribal practices in Mecca in the latter half of the nineteenth century. Not only can these practices be gleaned from the letter, but the two manuscript copies of historical texts that Snouck Hurgronje ordered from Mecca have also been preserved. This gives us unique insights into the theory and practice of text copying in Mecca in 1885–1889. Through an analysis of the content of this letter, which is here edited, translated and commented upon, it is now possible to understand what exactly was meant with the term 'Abū Šubbāk'.

1 Introduction

The history of European laid paper with watermarks in the world of Islam can be written from two different angles. One is that of the paper at its point of production in the paper mill. The sources for this part of paper history are not only the archives of paper manufacturers and the relevant rules and regulations of the countries in which they operated, but also accounts by the users of these papers. This research is part of the social and economic history of the paper industry. The other perspective is from the destination of the paper, from the place of arrival on the other side of the Mediterranean, to the Middle East,

 | DOI 10.1163/9789004360877_010

North and Sub-Saharan Africa, and beyond. The sources for our research on that side are books, documents and manuscripts, the ready products for which paper is the raw material. The most conspicuous feature in the paper on that other end of the line are the watermarks. With some good luck one can relate watermarks to data that are available on the production side and then draw conclusions. In theory the determination by the watermark of the origin and the age of paper becomes possible. In practice this is not so easy. Even with the numerous albums with watermarks that we have on the bookshelf, with the databases of watermarks that are now available, an individual identification of a watermark in the field to one in an album remains a rare event, if it happens at all. There are simply too many watermarks and variants of watermarks, especially in the nineteenth and early-twentieth centuries. To compound the difficulty, the mere presence of a certain watermark in the paper of a book or document is only an implicit message; it does not tell a story. At the end of the line there is usually no source available that is in any way comparable to the explicit stories told on the production end.

2 Introducing the Protagonists

Bits and pieces of such a story can be gleaned from the document that is the object of the present research. It is a letter in Arabic dated 10 Ǧumādā I 1303 (14 February 1886) that was sent by Raden Aboe Bakar Djajadiningrat (1854–1914), a Bantenese employee of the Dutch consulate in Jeddah,[1] to Pieter Nicolaas van der Chijs (d. 1889) [Fig. 8.1].[2] The latter, a Dutchman, was both consul of Sweden and the acting vice-consul in the Dutch consulate in Jeddah. He was also the agent of a number of shipping companies that were active in the Red Sea and beyond. Van der Chijs was the permanent Dutch consular representative in Jeddah. The Dutch consulate in Jeddah was administered from Batavia (Jakarta) and ultimately was the responsibility of the Dutch Ministry of Colonial Affairs, not that of Foreign Affairs. The consulate had been founded in

1 See on him M. Laffan, "Raden Aboe Bakar. An introductory note concerning Snouck Hurgronje's informant in Jeddah (1884–1912)", 1999, 517–542; id., "Writing from the colonial margin: The letters of Aboe Bakar Djajadiningrat to Christiaan Snouck Hurgronje", November 2003, 356–380.

2 Hardly any research has been conducted on this interesting personality. The few materials in the Snouck Hurgronje archive in the Library of the University of Leiden that concern Van der Chijs only give fragmentary moments in a life that would deserve a well-documented biography.

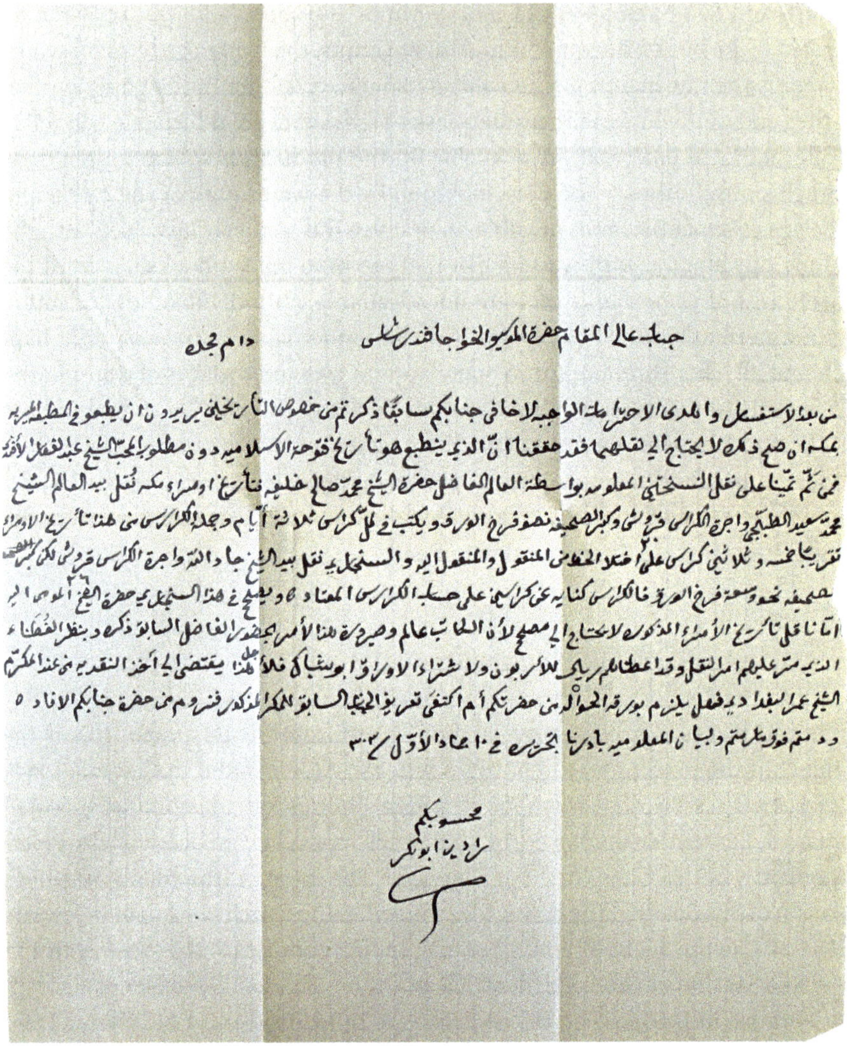

FIGURE 8.1 *The letter of Raden Aboe Bakar Djajadiningrat to consul P. N. van der Chijs in Jeddah, [Mecca] 10 Ǧumādā I 1303 (14 February 1886).* MS *Leiden, Or. 6977 A*

1869[3] and primarily served the Dutch-Indian government vis-à-vis its Muslim pilgrims. Pilgrims were monitored on pan-Islamic leanings and anti-Dutch activities, and at the same time they were protected against the exactions of the

3 U. Freitag, "Helpless Representatives of the Great Powers? Western Consuls in Jeddah, 1830s to 1914", September 2012, 357–381, 376, note 19, where the year 1872 is also given for the establishment of the Dutch consulate in Jeddah.

pilgrim guides. The often desperate health situation of the Hijaz was another point of concern for the consular representatives. In the 1920s and 1930s the Dutch consulate even had a branch office in Mecca itself.[4] After the Dutch had signed for Indonesia's independence, on 27 December 1949, the consulate building at that time was transferred to Indonesia. The Dutch consuls were usually only *en poste* in Jeddah during the pilgrimage season. Out of season they might go on long holidays to healthier places than Jeddah. Because of the unpleasant climate they never took their wives and families there.

That made the intrepid P. N. van der Chijs the permanent caretaker of Dutch interests in Jeddah. Raden Aboe Bakar was working under his orders and responsibility. This, however, was not their relationship in what concerns the document that is analysed hereunder. The third interested party was Christiaan Snouck Hurgronje (1857–1936),[5] who in 1886 and while in Leiden was writing the first volume of his book on Mecca.[6] He is mentioned in the letter by his Muslim name, ʿAbd al-Ġaffār Efendi. He had been expelled from Mecca by the Turkish authorities in the early days of August 1885 after a stay of slightly over five months in the Holy City. Because of his sudden departure he had been unable to bring back his notes and study materials. It is doubtful whether at that moment he already had in mind to write a monograph on Mecca, but while he was writing it later on, new documentary needs arose and developed.[7] His contact in Jeddah was Van der Chijs, whom he had befriended while he lived in Jeddah in 1884 and 1885. For materials that had to be provided directly from Mecca, Van der Chijs had to rely on his local informants, of whom Raden Aboe Bakar was probably the best. He was no stranger. He had been Snouck Hurgronje's companion on his journey from Jeddah to Mecca on 21 February 1885, and initially Snouck Hurgronje had lived in Raden Aboe Bakar's house in Mecca. The letter that is introduced hereunder must be placed within the context of Snouck Hurgronje's endeavours to obtain historical sources for the book that he was writing after his return to the Netherlands.

4 A. Patah, *De medische zijde van de bedevaart naar Mekka*, 1935, 55, gives a photograph of the pharmacy and the outpatients' clinic attached to this Meccan branch office, where all employees were Indonesian Muslims. The author worked there in 1926–1933. His book is his PhD thesis, which he defended in Leiden in 1935.

5 See on his life and work J. J. Witkam, "Christiaan Snouck Hurgronje", 2012, 148–154.

6 C. Snouck Hurgronje, *Mekka*. Vol. 1: *Die Stadt und ihre Herren*, 1888.

7 For a full account of the genesis of Snouck Hurgronje's *Mekka*, see my introduction in: Christiaan Snouck Hurgronje, *Mekka in de tweede helft van de negentiende eeuw. Schetsen uit het dagelijks leven*, 2007, 1–184, which makes abundant use of the correspondence between Snouck Hurgronje and Van der Chijs.

3 An Overview of the Source Material

Snouck Hurgronje was particularly in need of recent histories of Mecca, and more in particular of what he saw as the two most important sources, the *Manāʾiḥ al-Karam* by al-Sinǧārī, written in 1095/1684, and the *Ḫulāṣat al-Kalām* by Aḥmad Zaynī Daḥlān, which remained unfinished in 1303/1885. He knew of the existence of important copies in private hands in Mecca and he had asked Van der Chijs to procure for him copies of these. This implied that local copyists had to be employed for the speedy production of these texts that were later sent by Van der Chijs to Leiden and that were used by Snouck Hurgronje.[8] After Snouck Hurgronje's death in 1936 these materials were part of the huge Snouck Hurgronje bequest that entered Leiden's University Library and where they are still kept today.

a. *Manāʾiḥ al-Karam bi-Aḫbār Makka wa-al-Ḥaram* by al-Sinǧārī (wrote in 1095/1684; Brockelmann, *GAL* G II, 383–384, gives no more names than the *nisba*).

The author, who at Snouck Hurgronje's time was virtually unknown in Europe, came from a family which had established itself in Mecca over several generations and who had had friendly relations with the ruling *Šarīf*s.[9]

Presently two manuscripts of al-Sinǧārī's *Manāʾiḥ al-Karam* are preserved in the Leiden library. One is a 'very bad copy' of Sayyid Aḥmad b. Zaynī Daḥlān's copy. It is now registered in the Leiden library as Or. 7019.[10] This manuscript has no colophon and contains no details about the copying. On the basis of the letter from Raden Aboe Bakar to Van der Chijs we may assume that it was copied in 1303 or 1304 (1886 or 1887) by Šayḫ Ǧād Allāh and that the collations and corrections were made by Muḥammad Ṣāliḥ Ḫalīfa. The volume is made of European laid paper. It has 456 pp. of a large format (c. 34 × 24 cm, text area c. 24 × 13 cm), with 37 lines of text in *Naskh* script in black ink to the page. It is evidently the 'large copy' that is mentioned by Raden Aboe Bakar in his letter. Because of the large format, the sheets show the watermarks and countermarks in the central part of the page. The paper contains the watermark with the faced crescent in a shield (p. 456, which is a blank page and which therefore shows a clear image of the watermark). This watermark is what was commonly known in the Middle East as the Abū Šubbāk ('the man in the window')

8 He mentions these extensively in the preface to the first vol. of *Mekka*, 1888.

9 See on him C. Snouck Hurgronje, *Mekka*, 1888, vol. 1, xv ff.

10 P. Voorhoeve, *Handlist of Arabic manuscripts in the library of the University of Leiden and other collections in the Netherlands*, 1980, 2nd ed., 182.

logo, and that is what it shows, hence its name [Fig. 8.2]. Sometimes the image of Abū Šubbāk is combined with the text Abū Šubbāk.[11] In an unnumbered leaf between the flyleaf and the first leaf with text of this volume is the countermark, written twice, first as 'AG', then as 'Andrea Galvani | Pordenone' [Fig. 8.3]. The volume is bound in a half-leather European binding with pasted boards, and it has gilded ornaments and text on the spine. The volume consists of quires of five sheets (= 10 leaves = 20 pages), as is usual. According to the letter of Raden Aboe Bakar the copyist was not a learned person, and therefore a learned corrector was employed to indicate a better reading. For example, on p. 60, where *man qatalahu* in the text gets a marginal conjecture (*laʿallahu*) in a different hand: *man qātalahu*, which in the context is more suitable. The corrector has also written collation notes in the margin, e.g. on p. 58 *balaġa*.

The other manuscript of the *Manāʾiḥ al-Karam* is registered in the Leiden library as Or. 7018.[12] This manuscript was also copied for Snouck Hurgronje from an exemplar in Mecca, possibly the same copy of the text in Daḥlān's possession. According to the colophon on f. 595a it was copied on Friday 20 Rabīʿ II 1307 (14 December 1889) by al-Ṭibbəġī Muḥammad Saʿīd b. Muḥammad b. Sulaymān, the same scribe who also copied MS Leiden Or. 6977. It has 595 ff. plus 4 unnumbered leaves (24.8 × 17.3 cm, text area 19 × 9.5 cm) with 23 lines in *Naskh* script in black ink with rubrication to the page. Near the end of the volume the script becomes larger and the pages are filled more easily (e.g. ff. 581b–582a). As the copyist was paid per quire, this was to his advantage. The quires consist of five sheets (= 10 leaves = 20 pages). On the first recto page of the quires is a quire mark consisting of the letter *Kāf* with a figure, e.g. *Kāf* 2 (f. 11a), for *Kurrāsa* 2, etc. The volume consists of European laid paper with a faced crescent in a shield as watermark, the *Abū Šubbāk* watermark. As the paper size is smaller than that of Or. 7019, the watermark and the countermark can only partly be seen in the inner margins of the pages (watermark upper half on f. 1a, lower half on f. 3a; countermark upper half on f. 2a, lower half on f. 4a, etc.). A countermark in the outer margin occurs as well (f. 529). The volume has a half-parchment European binding, with author and title written in Latin script in black ink on the spine.

b. *Ḫulāṣat al-Kalām fī Bayān Umarāʾ al-Balad al-Ḥarām*, a history of the rulers of Mecca by Sayyid Aḥmad b. Zaynī Daḥlān (d. 1304/1886, Brockelmann, *GAL* G II, 499; S II, 810–811). This work, on a permanent update of which the

---

11 See the illustration in A. Regourd, "Introduction to the Codicology of the Collection", 2014, li.

12 P. Voorhoeve, *Handlist of Arabic manuscripts*, 1980, 182.

FIGURE 8.2 *Watermark* 'Abū Šubbāk' *in paper used for al-Sinǧārī's* Manā'iḥ al-Karam, *a manuscript copied in Mecca, probably in 1304/1887. MS Leiden, Or. 7019, unnumbered leaf before p. 456*

author worked till almost the end of his life, treats the history of Mecca's rulers up to 1303/1885. The older period of Mecca's history in it is largely modelled after and based on al-Sinǧārī's *Manā'iḥ al-Karam*. The Leiden manuscript of the *Ḫulāṣat al-Kalām* is a copy of the author's own copy. The text in the Leiden MS is, apart from the colophon, identical to that of the edition published in

FIGURE 8.3 *Countermarks 'A. G.' and 'Andrea Galvani | Pordenone' in paper used for al-Sinǧārī's* Manā'iḥ al-Karam, *a manuscript copied in Mecca, probably in 1304/1887. MS Leiden, Or. 7019, unnumbered leaf before p. 1*

Cairo in 1305/1888. It is now registered as MS Leiden Or. 6977.[13] It is a volume on European laid paper of 608 pp. of c. 24.3 × 17 cm and with a text area of c. 18.7 × 9.3 cm. According to the colophon on p. 608 it was copied on Tuesday 15 Ǧumādā I 1304 (9 February 1887) by al-Ṭibbəǧī Muḥammad Saʿīd b. Muḥammad b. Sulaymān, who later also copied Or. 7018, one of the two copies of *Manā'iḥ al-Karam* by al-Sinǧārī in the Leiden library.

Several watermarks can be observed in the paper on which this text was copied: a shield with a crescent with face (e.g. pp. 604–605, 607), which because of the size (and number of foldings of the paper) can only be seen in halves in the inner margins (e.g. pp. 583, 585: lower half; pp. 597, 595: upper half [Figs. 8.4, 8.5]). The countermark is 'Andrea Galvani | Pordenone' (e.g. pp. 509, 539 [Fig. 8.6]), or one can see the watermark together with the countermark (e.g. in p. 8). Another paper, with a double-headed eagle as watermark, is also

13 P. Voorhoeve, *Handlist of Arabic manuscripts*, 1980, 161.

FIGURE 8.4 *Watermark* 'Abū Šubbāk' (*upper half*), *in paper used for Daḥlān's* Ḫulāṣat al-Kalām, *a manuscript copied in Mecca, Tuesday 15 Ǧumādā I 1304* (*9 February 1887*). *MS Leiden, Or. 6977, p. 605*

FIGURE 8.5 *Watermark* 'Abū Šubbāk' (*lower half*), *in paper used for Daḥlān's* Ḫulāṣat al-Kalām, *a manuscript copied in Mecca, Tuesday 15 Ǧumādā I 1304* (*9 February 1887*). *MS Leiden, Or. 6977, p. 607*

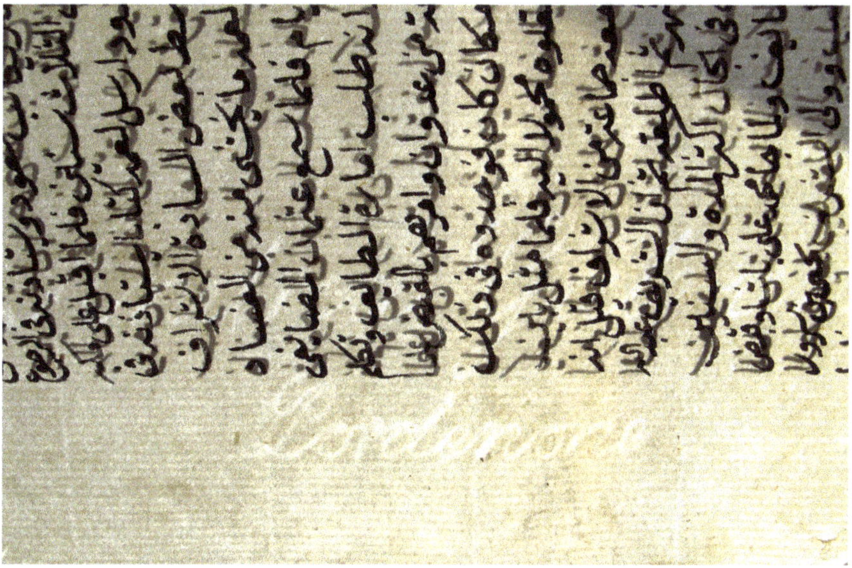

FIGURE 8.6 *Countermark 'Andrea Galvani | Pordenone', in paper used for Daḥlān's* Ḫulāṣat al-Kalām, *a manuscript copied in Mecca, Tuesday 15 Ǧumādā I 1304 (9 February 1887).* MS *Leiden, Or. 6977, p. 509*

used (e.g. pp. 585/586–596/595 and pp. 587/588–598/597 [Fig. 8.7]), but most watermarks show the faced crescent in the shield, 'Abū Šubbāk', 'the man in the window'. The quires consist, as usual, of five sheets (= 10 leaves = 20 pages).

Snouck Hurgronje's study notes on Daḥlān's *Ḫulāṣat al-Kalām* are preserved in the form of six exercise books in the Leiden library, where they are registered as Or. 7110. Aḥmad b. Zaynī Daḥlān, the author of the *Ḫulāṣat al-Kalām*, was the mufti of the Shāfi'ite *Maḏhab* in Mecca, and he is mentioned at several instances in the book *Mekka* and in the relevant correspondence. Snouck Hurgronje knew him personally, had visited him in Mecca on several occasions, and upon Daḥlān's demise in 1886 he wrote a long obituary.[14]

A loose insert of one leaf of paper (26 × 21 cm) in MS Leiden Or. 6977 with Arabic text on one side only is the letter from Raden Aboe Bakar to Van der Chijs of 10 Ǧumādā I 1303 (14 February 1886), which is treated hereafter. It is now registered in the Leiden library as Or. 6977 A.

14 "Een Rector der Mekkaansche Universiteit", 1887, 344–404. (*Verspreide Geschriften*, 1923, vol. 3, 65–122).

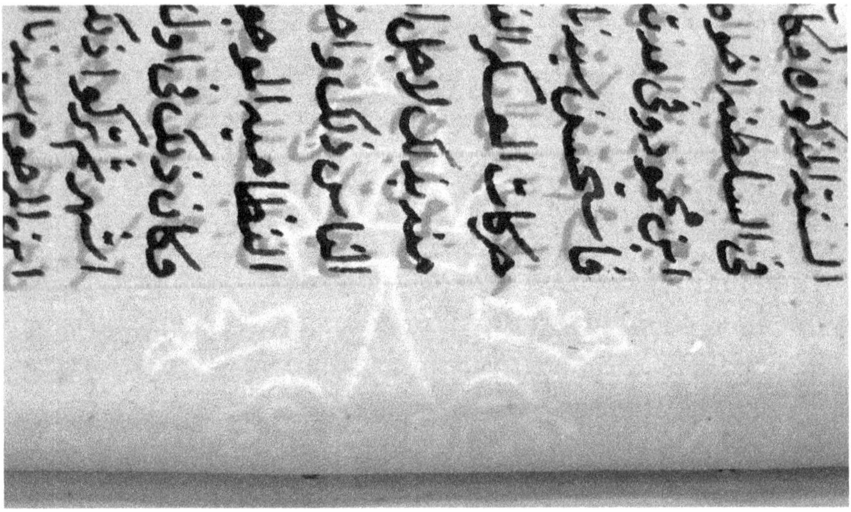

FIGURE 8.7 *Watermark of double-headed eagle (upper part only) in paper used for Daḥlān's* Ḫulāṣat al-Kalām, *a manuscript copied in Mecca, Tuesday 15 Ǧumādā I 1304 (9 February 1887).* MS *Leiden, Or. 6977, p. 595*

Also added to the volume are some loose sheets, which Snouck Hurgronje received in the course of 1913 and which contain a survey of the history of Mecca over the period of 1301–1331 (1883–1913).

4 The letter from Raden Aboe Bakar Djajadiningrat to Pieter Nicolaas van der Chijs

In the following a transcript of the letter of Aboe Bakar (MS Leiden Or. 6977 A) is given. The Arabic, which shows some features of the spoken language and several orthographical irregularities, has been copied as found in the document. Twice a letter which was evidently forgotten by the writer of the letter has been added between < >.

Transcription of the Arabic

جناب عالى المقام حضرة الموسيو انخواجافندى فندر شايس دام مجده

من بعد الاستفسار واهدى الاحترامات الواجبه لا خافى جنابكم سابقاً ذكرتم
من خصوص التاريخين يريدون ان يطبعو فى المطبعه الميريه | بمكه ان صح ذلك

لا يحتاج الي نقلهما فقد حققنا انّ الذي ينطبع هو تاريخ فتوحة الاسلاميه دون
مطلوب المحب الشيخ عبد الغفار الأفندى | فمن ثم تمّينا على نقل النسختين
المعلومه بواسطة العالم الفاضل حضرة الشيخ محمد صالح خليفه فتأريخ اومراء
مكه نُقل بيد العالم الشيخ | محمد سعيد الطبّجى واجرة الكراس ٢٠ قروش وكبر
الصحيفه نصف فرخ الورق ويكتب فى كلّ كراس ثلاثة أيّام وجملة الكراريس
من هذا تأريخ الاومراء | تقريبا خمسه وثلاثين كراس على اختلا<ف> الخط من
المنقول والمنقول اليه والسنجاري نقل بيد الشيخ جاد الله واجرة الكراس ٢٦
قروش لكن كبر | الصحيفه نحو وسعة فرخ الورق فالكراس كأيه عن كراسين
على حساب الكراريس المعتادة ويصلح فى هذا السنجاري حضرة الشيخ المومى
اليه | امّا ناقل تأريخ الأُمراء المذكوره لا يحتاج الي مصلح لأن الكاتب عالم
وصيرورة هذا الأمر بحضور الفاضل السابق ذكره وبنظر الفُطَناء | الذي مرّ عليهم
امر النقل وقد اعطناهم ٥ ريال للأربون ولاشتراء الاوراق ابو شباك فلأجل
هذا يقتضى الي أخذ النقديه من عند المكرّم | الشيخ عمر البغدادي فهل يلزم
بورقه الحواله من حضرتكم أم اكتفَى تعريف الجناب السابق للمكر<م> المذكور
فنروم من حضرة جنابكم الافاده | ودمتم فوق ما رمتم ولبيان المعلوميه بادرنا
بتحريره فى ١٠ جماد الأوّل سنه ٣٠٣

محسوبكم
رادين ابو بكر

Translation

To the Right Honorable, the Man of Sublime Position, His Excellency, Monsieur Ḫawāǧa Efendi Van der Chijs, may his glory be permanent.

After having enquired [after your health] and after [having paid] the most proper obligatory respect, which does not remain hidden for Your Honour. Earlier you have told [me] concerning the [two volumes on] history that they wish to print at the Governmental Printing Press in Mecca, [that] if this is true, it is not necessary to copy these two texts. We have found out that the book that is printed is the 'History of the Islamic Conquests', which is not desired by our friend Šayḫ ʿAbd al-Ġaffār

Efendi. Therefore we have completed [the procedure] of the [preparation of the] copying of the two specified texts through the intermediary of the Excellent Scholar, His Excellency Šayḫ Muḥammad Ṣāliḥ Ḫalīfa. The 'History of the Rulers of Mecca' is copied by the scholar Šayḫ Muḥammad Saʿīd al-Ṭibbəǧī. The remuneration is twenty piaster per quire, and the size of the page is half a leaf. He writes one quire in three days. The total of the quires of this 'History of the Rulers' is approximately thirty-five quires, taking into account the difference in script between the original and the copy. The ['History' by] al-Sinǧārī is copied by Šayḫ Ǧād Allāh. The remuneration per quire is 26 piaster, but the size of the page is about a full leaf, and then a quire [of his work] is the same as two quires as counted in the case of normal quires. This ['History' of] al-Sinǧārī must be corrected by the above-mentioned Excellency the Šayḫ. As to the copyist of the above-mentioned 'History of the Rulers', he does not need a corrector because the copyist is a scholar, and the procedure in this case [will be] in the presence of the Excellent [Šayḫ] aforementioned, and under the supervision of intelligent people, who have seen the request for copying passing by them. We have given them five *riyāl* as a payment up-front and in order to buy sheets of paper, Abū Šubbāk. For that purpose it is necessary to make a withdrawal of money from the Noble Šayḫ 'Umar al-Baġdādī. So, is it necessary to do this with a cheque that can be drawn on Your Excellency or is it sufficient that Your Excellency brings this to the attention of the Noble Šayḫ? We ask instructions from Your Excellency.

May you have a long life, above what you wish for. As soon as we had the [above] information [from you], we have without delay [written this letter] on 10 Ǧumādā I of the year [1]303.

Your obedient Raden Aboe Bakar.

A few remarks on the language of the document are in place. The letter is written in Arabic, a language that both the writer and the addressee were apparently familiar with, but which is not the first language of either one of them. The Arabic of the letter shows all sorts of peculiarities. First, it is far from perfect in the orthography of written Arabic and the reader of the Arabic text can easily recognize instances of this. I leave these for what they are as they do not impair the understanding of the text. The writer's use of the Arabic article shows that he is not familiar with its intricacies. Sundanese, his mother tongue spoken in Banten, Western Java, does not have the article. An example of colloquial Arabic is the verbal form *tammēnā* (تمّينا) in line 4, which would have been *tamamnā* in the literary language. A word that baffled me at first was *li-al-urbūn* (للأربون) in line 9. *Urbūn* should probably be read as *ʿUrbūn*, 'advance payment'.[15] On the other hand, Raden Aboe Bakar is truly proficient in the use of epistolary titles. The latter half of his letter is incomprehensible if one has not carefully studied the titles that the writer uses in the first half.

5 Conclusions

Raden Aboe Bakar must have written this letter in Mecca and had it taken from there to P. N. van der Chijs in Jeddah. He was organizing the copying of manuscripts in Mecca on behalf of 'our friend Šayḫ ʿAbd al-Ġaffār Efendi', that is: Christiaan Snouck Hurgronje. In his letter, Raden Aboe Bakar reports on the developments of the preparation of the copying job and he asks for further instructions. He may have purposely omitted the mention of Mecca while dating and signing his letter for reasons of security. He was, after all, actively involved in the gathering of useful knowledge on behalf of the Dutch consular authorities, who as non-Muslims could not enter Mecca. Security may also have been the reason why he did not mention the name of the author and owner of one of the texts to be copied, the mufti of the Shāfiʿiyya and the *Šayḫ al-ʿUlamāʾ* in Mecca, Aḥmad b. Zaynī Daḥlān. In 1886, shortly before his death (in Muḥarram 1304/October 1886), in the conflict between the Grand-Šarīf of Mecca and the Turkish governor, Daḥlān chose the side of the Šarīf and followed him in exile to Medina.[16] Either manuscript to be copied was in the library of Daḥlān in Mecca. In that tense political situation, mentioning Daḥlān's name in writing may not have been a wise thing to do. Finally, if Raden Aboe Bakar had not been in Mecca, but in Jeddah, there would have been no need to write a letter, he could simply have told Van der Chijs personally.

The letter of Raden Aboe Bakar can be placed in a context. This becomes clear from Snouck Hurgronje's own publications on the sources which he used for the first volume of his book *Mekka*. Particular details on these sources and the circumstances under which they were copied can also be found in documents in the Snouck Hurgronje Archive in Leiden University Library. This is particularly the case with the correspondence between Snouck Hurgronje and Van der Chijs,[17] and with a letter from Muḥammad Ṣāliḥ Ḫalīfa, a professor

15 *Muʿǧam al-Rāʾid al-Luġawī*, an on line dictionary, s.v.

16 C. Snouck Hurgronje, "Een Rector der Mekkaansche Universiteit", 1887, 344.

17 Significant parts of this correspondence, which is almost entirely written in Dutch, have been edited by me in my introduction to Christiaan Snouck Hurgronje, *Mekka in de tweede helft van de negentiende eeuw*, 2007.

in Mecca's Great Mosque, to Snouck Hurgronje. That letter will be referred to hereafter. Raden Aboe Bakar's letter is special insofar as it was never deposited in the Snouck Hurgronje Archive. It remained unknown because it was kept as a loosely inserted leaf in one of the manuscripts it directly concerned, now Or. 6977 in the Leiden library, which is Snouck Hurgronje's copy of Daḥlān's autograph of *Ḫulāṣat al-Kalām*.

The two originals to be copied were kept in the library of Mecca's *Šayḫ al-ʿUlamāʾ* Aḥmad b. Zaynī Daḥlān. Their owner may still have been in Mecca when Raden Aboe Bakar wrote his letter to Van der Chijs on 10 Ǧumādā I 1303 (14 February 1886). If so, it cannot have been for long, as he was to join the Grand-Šarīf in Medina very soon. The following situation becomes clear from the letter. Raden Aboe Bakar had asked for assistance from a Meccan scholar, Šayḫ Muḥammad Ṣāliḥ Ḫalīfa,[18] in finding copyists for the copying job. The letter mentions the task of copying of two historical works. The first work is referred to as *Tārīḫ Umarāʾ Makka*, the 'History of the Rulers of Mecca'. This must be Aḥmad Zaynī Daḥlān's work *Ḫulāṣat al-Kalām*. As prospective copyist Šayḫ Muḥammad Saʿīd al-Ṭibbəǧī is mentioned, and he is indeed the copyist of MS Leiden Or. 6977 (colophon on p. 608). This copyist was also a scholar, according to Raden Aboe Bakar's letter, and therefore there was no need for a corrector who would inspect and correct this copyist's work. The second work to be copied is referred to in the letter as 'al-Sinǧārī', who is the author of *Manāʾiḥ al-Karam*. As prospective copyist a certain Šayḫ Ǧād Allāh is mentioned. He was not a scholar, Raden Aboe Bakr tells Van der Chijs, so after the copying a corrector had to go through his work. That was to be Šayḫ Muḥammad Ṣāliḥ Ḫalīfa himself, who acted as the intermediary for the two copying jobs. Šayḫ Ǧād Allāh would write on large format paper. This cannot but refer to the manuscript that is now MS Leiden Or. 7019. That manuscript is indeed written on large format. It seems to be incomplete at the end. There is no colophon, and neither copyist, nor collator is mentioned. It contains an almost negligible amount of marginal corrections. The copies should be written on high quality paper, which Raden Aboe Bakar indicates as 'Abū Šubbāk'. If we look in

18 Muḥammad Ṣāliḥ Khalīfa signs as '*al-Mudarris bi-al-Masǧid al-Ḥarām*', 'Professor in the Great Mosque of Mecca', under a letter to al-Shaykh ʿAbd al-Ġaffār al-Hūlanda (= Snouck Hurgronje) of 17 Q[aʿda] 1302 (28 August 1885), Snouck Hurgronje Archive, University Library of Leiden, Or. 8952 A: 510 (available online through the library's general catalogue). At the time Snouck Hurgronje was still in Jeddah. He left Jeddah by ship, possibly on 18 or 19 September 1885, since he arrived in Suez on 21 September. In the letter of Muḥammad Ṣāliḥ Khalīfa mention is made of Snouck Hurgronje's wishes to copy the 'History of the Rulers of Mecca' by Daḥlān and the 'History' by al-Sinǧārī (line 8).

the manuscripts themselves, we see the watermark of the faced crescent in a shield and the countermark of Andrea Galvani of Pordenone, the well-known Italian paper makers.[19] That is, therefore, what the name 'Abū Šubbāk' refers to in the present context. Mentioning the watermark is necessary, because it identifies the paper that should be used. As the price of the paper (five *riyāl*) was already given, mention of the watermark would prevent the copyist from purchasing a cheaper brand of paper.

The 'History of the Islamic Conquests', which is also mentioned in the letter, is yet another work by Daḥlān.[20] This work was not on the list of Snouck Hurgronje's desiderata from Van der Chijs as he had probably already obtained a copy of the printed edition, which had just appeared in two volumes.

Raden Aboe Bakar's letter gives us some insight into the speed of copying and the costs involved. Copyist Šayḫ Muḥammad Saʿīd al-Ṭibbəǧī would write three days for each quire of Daḥlān's *Ḫulāṣat al-Kalām*. The quires in MS Leiden Or. 6977 consist of five sheets (= 10 leaves = 20 pages). The pages in this manuscript have a text area of 18.7 × 9.3 cm = 173.91 cm² per page. The average number of words per line is 14, with 23 lines to the page. This gives an estimated number of 14 × 23 × 20 words = 6440 words per quire. To realize one such quire in three days on the surface available seems feasible, I would say. The copyist of Or. 6977 has maintained from the beginning (p. 1) till the end (p. 608) the same size of script, so that we can assume that he had 30 quires to copy, which he would do in 90 days, three months. The estimate of thirty-five quires that Raden Aboe Bakar mentioned in his letter was too high. The copyist would receive 20 piasters per quire, which sum did not include the cost of paper and the advance payment. For this the sum of five *riyāl* was available.

If, in Mecca at the time, a *riyāl* is the equivalent of twenty piasters, we can make the following tentative calculation for the total cost of the paper and other requisites, and the cost of the copying.[21] There is one uncertainty: it is not clear from Raden Aboe Bakar's letter how much 'Abū Šubbāk' paper could be purchased with five *riyāl*. If I understand the text well, the sum of five *riyāl*

---

19 See G. Ganzer, *Andrea Galvani (1797–1855). Cultura e industria nell' Ottocento a Pordenone*, 1994.

20 C. Snouck Hurgronje, in "Een Rector der Mekkaansche Universiteit", 1887, 353–370, gives a survey of the content of and extensive quotations from *Al-futūḥāt al-islāmiyya baʿda muḍīʾ al-futūḥāt al-nabawiyya*, which was published in two vols. in Mecca in 1302/1884.

21 The rate of twenty piaster to the riyal is not certain. A contemporary document (MS Leiden, Or. 7111, map 12), also from Raden Aboe Bakar, which contains a pricelist of books ordered by Snouck Hurgronje in the course of December 1885, gives the impression of a rate of 28 piaster to the riyal being used.

concerns the paper for both Or. 6977 and Or. 7019. These two manuscripts are of different size. The sheets of Or. 6977 are half the size of those of Or. 7019. I assume that the sheets of paper sold for Or. 7019 were folded once, and that these same sheets, when used for Or. 6977, were folded twice. This means that for the sum of five *riyāl* (= one hundred piasters) some 76 sheets (for Or. 6977 of 608 pp. half-size) + 114 sheets (for Or. 7019 of 455 pp. full-size) = 190 sheets in total, plus writing implements, could be purchased. If 35 quires of paper were purchased, according to the estimate in the letter, there would be a surplus of paper for five quires, which would probably go to the copyist as a remuneration in kind.[22]

The copying of Or. 6977 would cost, according to the estimate of Raden Aboe Bakar, the sum of 35 quires @ 20 piasters = 700 piasters. According to Aboe Bakar's estimate, the copying of Or. 7019 would cost 26 piaster per quire. The expected number of quires of Or. 7019 is not given by Aboe Bakar, but Or. 7019 actually consists of 23 quires. That would bring the expense for the copying of Or. 7019 to 23 quires @ 26 piaster = 598 piaster. The labour would cost a total of 700 + 598 = 1298 piasters, whereas the material including the (non-recoverable?) advance would cost 100 piasters. This gives a ratio of labour/material of roughly 13:1.

Financial arrangements for the making of the copies are mentioned by Raden Aboe Bakar as well. Already in his letter to Snouck Hurgronje of 17 Ḏū al-Qaʿda 1302 (18 August 1885), Muḥammad Ṣāliḥ Ḫalīfa, who later coordinated the copying job, had mentioned the financial implications, but that was during the pilgrimage season: 'Even if we would give them [the copyists] one *riyāl* per quire, they would not even consider such a weak offer, and they would only begin to copy the books after the tenth of Muḥarram [1303 (19 October 1885)].'[23] Ultimately, the copying of Or. 6977 would indeed cost one *riyāl* (20 piaster) per quire. That arrangements with ʿUmar al-Baġdādī should be made for coming up with the necessary cash was also clear to Muḥammad Ṣāliḥ Ḫalīfa.[24] ʿUmar al-Baġdādī in Mecca would provide the cash money for the copying, which

22 In a totally different Islamic region the adagium is *al-Faḍl li-al-Ṣannāʿ* or *al-Faḍla li-al-Ṣannāʿ*, 'the surplus is for the maker'. See the anecdote in Ḥassāniyya Arabic from Senegal in MS Leiden Or. 14.055 E 4 (J. J. Witkam, *Catalogue of Arabic Manuscripts in the Library of the University of Leiden and other Collections in the Netherlands*, fasc. 1, 1983, 108–109).

23 Letter from Muḥammad Ṣāliḥ Khalīfa to Snouck Hurgronje of 17 Q[aʿda] 1302 (28 August 1885), lines 10–11. Snouck Hurgronje Archive, University Library of Leiden, Or. 8952 A: 510.

24 Letter from Muḥammad Ṣāliḥ Khalīfa to Snouck Hurgronje of 17 Q[aʿda] 1302 (28 August 1885), line 14.

would be refunded by Van der Chijs—and then by Snouck Hurgronje, we may assume. ʿUmar al-Baġdādī apparently had some kind of banking arrangement with Van der Chijs, who, if only for his intelligence work in Mecca, would constantly be in need of cash to be paid out in the Holy City. In March 1885 it had also been ʿUmar al-Baġdādī who provided Snouck Hurgronje with the sum of 150 Maria Theresia dollars for the purchase of an Ethiopian slave girl.[25]

About obtaining a copy of Daḥlān's as yet unpublished 'History of the Rulers of Mecca' Snouck Hurgronje wrote:

> The most important work of the Shaykh has not appeared in print and is not yet entirely completed. Either circumstance has to do with the nature of the book. It is a history of the rulers of Mecca, from Muḥammad to our time. In it the author continuously registered the most important political events of the Hijaz, of which he himself was a witness. He would have completed his work if he had been able to foresee his approaching demise. This history has been written in order to be read, but in the final part it contains all sorts of details about people who are still alive, notably from the family of the Sharīfs, that should not immediately be divulged in print. The manuscript was therefore only given for reading to a few friends of the author. Very few received permission to copy it, and if so, they were not allowed to copy the two or three final quires. During my stay in Mecca I was fortunate to become so well acquainted with the Shaykh that he give me the text for reading. Very soon I understood which invaluable and unique contributions Daḥlān's work contained to the history of the Holy City in the last two centuries. Immediately I tried to find learned and efficient copyists in order to have a copy made. Such copyists are very rare in Mecca, and when I was on the point of success, I suddenly had to leave the city for ever. Thanks above anything else to the friendly collaboration of our vice-consul, Mr. P. N. van der Chijs, I am now in the pleasant possession of a complete copy of the history of the rulers of Mecca.[26]

---

25 See for this the letter from Snouck Hurgronje to Van der Chijs, Mecca, 10 March 1885. Original in the Snouck Hurgronje Archive in Leiden University Library. My edition of the Dutch text of the letter is in: Christiaan Snouck Hurgronje, *Mekka in de tweede helft van de negentiende eeuw*, 2007, 97.

26 My translation from the Dutch passage in Snouck Hurgronje, "Een Rector der Mekkaansche Universiteit", 1887, 370–371. The manuscript is now MS Leiden Or. 6977. In the same article Snouck Hurgronje gives a long survey and extensive excerpts of this text. A shorter

Since about mid-July 1885, when still in Mecca (though not for long as he was to be expelled about two weeks later), Snouck Hurgronje had been able to borrow Daḥlān's 'History of the Rulers of Mecca', but for reading only. He decided to have a copy made secretly. On 26 July 1885 he wrote to Van der Chijs:

> Sometimes it is quite a job to get my desires here satisfied. At this moment I have in my house a large manuscript, which I borrowed from the *Shaykh* of the professors, *for reading only*. I would like to have a copy made of it in secret and quickly, but up till now I have not been able to find a *Kātib*, a 'copyist'. Most of them are, now that the 'season' has started, either working as *Muṭawwifs*, 'pilgrim guides', or they prefer to use their time by writing short letters for pilgrims against excessive sums of money, rather than by copying day after day a long manuscript.[27]

About his copy of al-Singārī's *Manā'iḥ al-Karam* (now MS Leiden Or. 7019) Snouck Hurgronje wrote: 'One of the best works on the more recent books of this nature, and that was highly appreciated by Daḥlān, is the History of a certain al-Singārī, which was written in the year 1095/1684. It seems to be unknown in Europe. I own a very bad copy of Sayyid Aḥmad Daḥlān's copy'.[28]

It is clear that Snouck Hurgronje was not happy with the copy of al-Singārī's *Manā'iḥ al-Karam* that he had ordered to be made from the text in Daḥlān's library. The reason may have been the fact that its copyist, Šayḫ Ǧād Allāh, was not a scholar and that the promised collation by Muḥammad Ṣāliḥ Ḫalīfa was next to non-existent. He must have decided that a new copy had to be made, and this time by the expert copyist Šayḫ Muḥammad Sa'īd al-Ṭibbəǧī, who had proven his value with the copying of Daḥlān's *Ḫulāṣat al-Kalām*. It is not clear which exemplar was used, and it is possible that this was just a better version of the same text in Daḥlān's library that was now more expertly

parallel text is in C. Snouck Hurgronje, *Mekka in the Latter Part of the 19th Century. Daily Life, Customs and Learning. The Moslims of the East-Indian Archipelago*, 2007, 178.

27 Letter from Snouck Hurgronje to Van der Chijs, Mecca, 26 July 1885. Original in the Snouck Hurgronje Archive in Leiden University Library; here translated from my edition of the letter in: Christiaan Snouck Hurgronje, *Mekka in de tweede helft van de negentiende eeuw*, 2007, 123. Snouck Hurgronje has the information about reasons of the scarcity of copyists in the pilgrimage season from the letter of Muḥammad Ṣāliḥ Khalīfa of 17 Q[a'da] 1302 (28 August 1885), line 9. Snouck Hurgronje Archive, University Library of Leiden, Or. 8952 A: 510.

28 My translation from C. Snouck Hurgronje, "Een Rector der Mekkaansche Universiteit", 1887, 371–372.

copied. The copying of that copy (now MS Leiden Or. 7018) was completed on Friday 20 Rabīʿ II 1307 (14 December 1889), too late to play any role in the composition of the first volume of *Mekka* that had appeared in print more than a year earlier.

Bibliography

Freitag, Ulrike. September 2012. "Helpless Representatives of the Great Powers? Western Consuls in Jeddah, 1830s to 1914", *The Journal of Imperial and Commonwealth History* 40/3, 357–381.

Ganzer, Gilberto. 1994. *Andrea Galvani (1797–1855). Cultura e industria nell' Ottocento a Pordenone*, Pordenone, Ed. Studio Tesi.

Laffan, Michael. 1999. "Raden Aboe Bakar. An introductory note concerning Snouck Hurgronje's informant in Jeddah (1884–1912)", *Bijdragen tot de Taal-, Land- en Volkenkunde* 155/4, 517–542.

Laffan, Michael. November 2003. "Writing from the colonial margin: The letters of Aboe Bakar Djajadiningrat to Christiaan Snouck Hurgronje", *Indonesia and the Malay World* 31/91, 356–380.

Muʿǧam al-Rāʾid al-Luġawī, https://al3rbe.com/dictionary/111115/ (accessed on February 1, 2015).

Patah, Abdoel. 1935. *De medische zijde van de bedevaart naar Mekka*, Leiden, Luctor et Emergo.

Regourd, Anne. 2014. "Introduction to the Codicology of the Collection", in: Alessandro Gori & coll., *A Handlist of the Manuscripts in the Institute of Ethiopian Studies. Vol. 2: The Arabic Materials of the Ethiopian Islamic Tradition*, Eugene (Or.), Pickwick Publications.

Snouck Hurgronje, C. 1887. "Een Rector der Mekkaansche Universiteit", *Bijdragen tot de taal-, land- en volkenkunde* 36, 344–404. And: *Verspreide Geschriften*, Bonn/Leipzig, 1923, vol. 3, 65–122.

Snouck Hurgronje, C. 1888. *Mekka*. Vol. 1: *Die Stadt und ihre Herren*, The Hague, Martinus Nijhoff.

Snouck Hurgronje, C. 2007. *Mekka in the Latter Part of the 19th Century. Daily Life, Customs and Learning. The Moslims of the East-Indian Archipelago*, Leiden, E.J. Brill.

Voorhoeve, Petrus. 1980. *Handlist of Arabic manuscripts in the library of the University of Leiden and other collections in the Netherlands*, The Hague/Boston, Leiden University Press [2nd ed.].

Witkam, Jan Just. 1983. *Catalogue of Arabic Manuscripts in the Library of the University of Leiden and other Collections in the Netherlands*, fasc. 1.

Witkam, Jan Just. 2007. Introduction, in: Christiaan Snouck Hurgronje, *Mekka in de tweede helft van de negentiende eeuw. Schetsen uit het dagelijks leven*, Vertaald en ingeleid door Jan Just Witkam. Amsterdam/Antwerpen, Uitgeverij Atlas, 1–184.

Witkam, Jan Just. 2012. "Christiaan Snouck Hurgronje", in: Coeli Fitzpatrick & Dwayne A. Tunstall (eds.), *Orientalist Writers* (= *Dictionary of Literary Biography*, vol. 366), Detroit/New York/San Francisco/New Haven (Conn.)/Waterville/Maine/London, Gale, 148–154.

CHAPTER 9

Collection of Persian *farmāns* on Russian Paper in the National Library of Russia (two first decades of the 19th century)

Olga Yastrebova

Résumé

Cette étude décrit et analyse une collection de *farmāns* persans des deux premières décennies du xixe s. Ils ont été écrits au nom du Shah iranien Fatḥ ʿAlī à plusieurs reprises durant la guerre russo-persane qui s'est déroulée dans le Caucase en 1804–1813. Le support des documents est du papier de fabrication russe produit par trois moulins.

1 Introduction

Paper manufactured in the Russian Empire was exported to Iran in the Qājār era, a fact that is confirmed by manuscripts, handwritten documents and lithographed books produced in Iran in the late 18th and 19th centuries with the use of Russian paper as support material. Efforts were made on the part of Iranians towards importing the technology as well, as reflected in recently published documents from 1850s regarding the purchase in Russia of a hand-operating paper mill, its delivery to Iran, and the finding and bringing in of a technical specialist to run it and teach local personnel.[1] The history of the Russian paper import in Iran has not yet been studied comprehensively, although the history of paper production in Russia is a well-studied field, and so is the general history of the Russo-Iranian commerce. This topic still awaits a researcher who would study the general statistics and their dynamics, as well as particulars, such as, for example, personalities involved in the trade, paper mills whose products found their way to Iran, spheres of the usage of Russian paper, etc.

The purpose of this article is to present and analyze a small collection of the Persian *farmāns* issued by Fatḥ ʿAlī Šāh and dealing with dramatic events that took place immediately before and in the course of the Russo-Persian war in the Caucasus of 1804–1813. The earliest document is dated 19 Rağab 1218/5

1 See R. Farāsatī, "Vārid kardan-i asbāb-i kārkhāne-yi kāghazsāzī az Rūsiye dar dawre-yi Nāṣirī", 1390/2012.

 | DOI 10.1163/9789004360877_011

November 1803, the last one—Ǧumādā I 1224/June–July 1809. The documents of this collection are united by the events reflected in them, by the time, place and circumstances of their appearance, and by the history of their acquisition by the National Library of Russia. Their physical execution is also almost identical and Russian paper of blue color serves as support for the most of them (with one exception). We shall discuss the documents against their historical background and analyze the watermarks found on their paper. Of course this small corpus is by no means extensive enough to draw any general conclusions from regarding the history of Russian paper in Iran even during this limited period of time. Nevertheless, we hope that the results of the study of this small collection will be of use for further researchers in the field.

2 The Collection

Twenty-four original Persian documents under examination here were part of the donation, consisting of 31 Persian *farmāns* in total, by the Great Prince Mihail Nikolaevič (the Governor-General of the Caucasus in 1863–1869) to the Library (then named Imperial Public Library) in 1869. This was his second such donation, the first having taken place three years earlier in 1866 and including ten charters in Georgian, nine documents in Persian, one in Turkish and one in Arabic, all originating from the Georgian-Imeretian Synodal Office (*Gruzinsko-Imeretinskaâ sinodal'naâ kontora*). Its collection was started in 1818 by the Exarch of Georgia,[2] Feofilakt (Theophilactus) (1765–1821), who gathered materials scattered among ancient churches and monasteries. The Georgian documents of this earlier party were transferred to the Georgian Soviet Socialist Republic in 1923, while those in the Arabic script, dating from the 17th and 18th centuries, remained within the walls of the National Library.

The texts of most of the Persian and Georgian documents from both collections donated by Mihail Nikolaevič were published in the fundamental work "Acts collected by the Caucasian Archaeographic Commission" edited by the Russian archaeographer Adolf Petrovič Berže (1828–1886).[3] In this publication their texts are represented in a slightly abridged form, with the omission of introductory parts.

2 The Georgian exarchate of the Russian Orthodox Church, established in 1811, existed until 1920 when Authocephalos Georgian Orthodox church was restored.

3 *Akty, sobrannye kavkazskoj arheografičeskoj kommissiej*, 1866–1904; the 12 vols. published between 1866 and 1904 (the last two with the editorship of Dmitrij Kobâkov) introduced valuable sources on the history of the Caucasian people in a number of languages.

3 Historical Context

The twenty-four Fatḥ ʿAlī Šāh's *farmāns* date from the years 1803–1805 and 1809, and they are now part of the Collection no. 1236 "Persian Documents", items no. 14–37.[4] We shall briefly describe their contents pointing at the events that brought about their issue.

Towards the end of the 18th century the rivalry between Russia and Iran in the Caucasus was becoming more and more intense, as Russia was striving to spread its power over this region. The kingdom of Kartli and Kakheti (Eastern Georgia) had been a Russian protectorate since 1783, after the signing of the Treaty of Georgievsk, and until 1801, when after the death of King Giorgi XII (1746–1800, ruled 1798–1800) it was incorporated into the Russian Empire and Georgian monarchy was abolished. General Prince Pavel Cicianov (1754–1806), whose personality and achievements received mixed assessments from both contemporaries and later historians, was appointed commander-in-chief in Georgia in 1803.[5]

Fatḥ ʿAlī Šāh's *farmān* (Pers. Doc.14) dated 19 Rağab 1218/5 November 1803 is addressed to Muḥammad Ḥusayn Bīg, Nasīb Bīg, Muṣṭafā Bīg, and other commanders and elders of Zāgām territory in Qarabagh inhabited by the Šams al-dīnlū (Šamšadil) tribe, a branch of the Qizilbāš Ḏulqadar tribe. The Shah informs the addressees that an army of 20,000 warriors lead by Pīr-Qulī Khān Qājār and Georgian Prince Alexander (Aliksandr Mīrzā)[6] is being sent to Ganjeh, and orders the elders with their militia to join Javād Khān, the *beglerbeg* of Ganjeh, in his struggle against the Russian forces.[7] In January 1804 united Russian and Georgian armies under Cicianov's command assaulted Ganjeh and after several weeks of siege occupied the city. The fortress was of paramount strategic importance, as Russian control over it could secure the eastern borders of Georgia.

---

4 See their description in: O. M. Âstrebova, *Persidskie i tadžikskie dokumenty v Otdele rukopisej Rossijskoj Natsional'noj biblioteki: Sistematicheskij katalog*, 1999, 27–60.

5 See his characteristic in: M. Atkin, *Russia and Iran: 1780–1828*, 1980, 72–75.

6 Alexander (1770–1844), one of Erekle II's sons in his third marriage with Daredjan Dadiani and step-brother of the last Georgian King Giorgi XII. After his father's death he and his other brothers rejected Russian dominion over Georgia. Alexander joined Omar, Khan of the Avars, in his assault on Kakheti in 1800. Having been defeated, the Prince fled to Qarabagh and Daghestan, and then to Iran where he actively participated in the Persian campaigns in Georgia. See: G. Bournoutian, "Alexander, Prince", *Encyclopaedia Iranica*, I/8; D. M. Lang, *The last years of the Georgian monarchy, 1658–1832*, 1957, 233, 238, 256.

7 See *Akty*, 1866–1904, II, 803, no. 1653.

Meanwhile the Western Georgian kingdom of Imeretia was forced to nominally submit to Russia in April 1804, though its ruler King Solomon II of the Bagratid dynasty (1772–1815, ruled 1789–1790, 1792–1810) was looking towards Persia and Turkey, hoping that they would help Imeretia to keep its independence. Fatḥ ʿAlī Šāh in his turn tried to have Solomon II on his side and addressed him with the *farmān* (Pers. Doc.15; Ḏū al-ḥiǧǧa 1218/March–April 1804) informing him that Aliksandr Mīrzā and Taymūraṯ Mīrzā[8] had been sent to join ʿAbbās Mīrzā[9] and his army. The Shah promised to come to the theater of war in person after the month of Ṣafar[10] and ordered Solomon (or Sulaymān, as he is named in the text) to gather his forces and await the Persian army being ready to join it.

Young ʿAbbās Mīrzā and his regiments collided with the Russian army at Echmiadzin while Cicianov was leading his forces towards Yerevan.[11] Cicianov continued towards Yerevan, where he again was attacked by the joined forces of ʿAbbās Mīrzā and Aliksandr Mīrzā. This attack was repelled, but Cicianov was not able to conquer the citadel of Yerevan and had to besiege it. The siege lasted from July to September, Russian troops found themselves surrounded and cut off from supply, suffering heavy losses and finally lifting the siege and returning to Tiflis.[12] The Yerevan campaign of 1804 became the beginning of the officially proclaimed Russo-Persian war that lasted till 1813.

Inspired by this success Fatḥ ʿAlī Šāh sent a number of *farmān*s to Solomon II (Pers. Doc.17), the people of Tiflis (Pers. Doc.19), the khans of Kabarda, and the Chechens and Ossetians (Pers. Doc.22) describing this military achievement and instigating the addressees to further actions against the

8 Tejmuraz Bagrationi (1782–1846), the fourth and youngest son of Giorgi XII, Prince Aleksandr's nephew. He also escaped to Persia, where during the war he commanded Persian artillery. In 1810 Tejmuraz voluntarily submitted to the Russian authorities and thereafter lived the life of a scholar in Saint Petersburg. He devoted himself to the study of Georgian history and literature, was elected the honorary member of the Russian Academy of Sciences and is credited with the foundation of the St. Petersburg School of Kartvelian studies. D. M. Lang, *The last days*, 1957, 255–256; "Bagrationi, Tejmuraz Georgievič", 2005.

9 ʿAbbās Mīrzā (1789–1833)—the Governor of Azerbaijan and Commander of the Iranian army in the wars in the Caucasus, the proclaimed heir to Fatḥ ʿAlī Šāh. When the war broke he was only 15 years old.

10 See *Akty*, 1866–1904, II, 805, no. 1656; advance of the Persian armies lead by Prince Alexander, ʿAbbās Mīrzā and the Shah himself, is also mentioned in nos. 1658, 1660–1662 etc.

11 See M. Atkin, *Russia and Iran*, 1980, 120.

12 M. Atkin, *Russia and Iran*, 1980, 76–77, 120–121.

Russians. Ossetian, Qarakalhani and Circassian hostages kept in Tehran were released and sent to their homeland as a token of Shah's gratitude for the support by their peoples (Pers. Doc.23, 24). However a month later, in December-January 1805, the Shah demanded that new hostages be sent by the Qarakalhanis (Pers. Doc.33), Circassians (Pers. Doc.27, 30), Tagaurs (Pers. Doc.28), Qurtauls (Pers. Doc.29), Avars (Pers. Doc.31), and Chechens (Per. Doc.32).

Farnavāz Mīrzā,[13] the leader of the Georgian rebellion against Russian dominion that broke out in the summer of 1804 in the mountainous parts of Kartli, also received several edicts from Fatḥ 'Alī Šāh with the news of Cicianov's failure at Yerevan (Pers. Doc.20, 35), as well as orders to gather militia in Qarakalhan and join Pīr-Qulī Khān and Aliksandr Mīrzā in their planned assault of Tiflis (Pers. Doc.21). Farnavāz Mīrzā also received information about Fatḥ 'Alī Šāh's intention to attack Georgia after the celebration of Nawruz in 1805 (Pers. Doc.24). This latter plan was also announced to the Prince Tahmāsb Bīg Gurjī and to a certain Muḥarram Bīg (Pers. Doc.26) and Muḥammad Ḥasan (Pers. Doc.34).

After Cicianov's assassination in Baku in 1806 General Field Marshall Ivan Gudovič (1741–1820) was appointed as his successor. Simultaneous wars with Napoleon in Europe and with Turkey in the Caucasus then made Russia seek a temporary truce with Persia. Four undated letters addressed by 'Abbās Mīrzā to Gudovič discuss the terms of a truce and mention the mediation of France in the matter (Pers. Doc.38, 39, 40 and 41). These letters, written on Oriental and *Tre lune* watermarked paper, are not included further into this survey.

Military actions resumed in 1808 when Gudovič occupied Nakhichevan and made another unsuccessful attempt to besiege and storm Yerevan, his failure costing him his position. During the last stage of the war the Russian armies in Georgia were commanded by General Alexander Tormasov (1752–1819). After decisive Russian victories at Lenkaran and Aslanduz the war resulted in the signing of the Gulistan Peace Treaty and the Persia's loss of all claims on the lands north of Kura and Aras, including Daghestan, Georgian lands (Kartli, Kakhetia, Megrelia, Imeretia, Guria, Abkhazia), Talesh and several khanates on the territory of modern Azerbaijan.

From this second phase of the war there are two *farmāns*, dated Rabī' al-āḫir 1224/May-June 1809 and Ǧumādā I 1224/June–July 1809. The first was sent

13 Parnaoz Bagrationi (1777–1852) was the youngest son of King Erekle II and brother of Prince Alexander. Together with his brother Ûlon he fled to Imeretia in 1801. Captured in 1804 by the Russian forces, he spent the rest of his life in Russia, first in exile in Voronež, and later in St. Petersburg. D. M. Lang, *The last years*, 1957, 233, 256; *Dvorânskie rody Rossijskoj imperii*, 1996, vol. 3, 70.

to Salīm Khān of Shakki, to confirm him in office. The second one, just as the earliest document in this collection, was issued for the elders of Šamšadil, informing them about the dispatch of the army of 50,000 soldiers under the command of the Qājār Prince Muḥammad ʿAlī Mīrzā.

4 The *farmāns*

All the *farmāns* of Fatḥ ʿAlī Šāh in this collection bear the imprint of his seal of rectangular shape with a "small cap", dated 1217 (1802–1803) with the following legend: "العزة لله ⭑ گرفت خاتم شاهی ز قدرت ازلی ⭑ قرار در کف شاه زمانه فتحعلی ۱۲۱۷"—"The glory belongs to God! The royal signet, because of the eternal power, has established itself on the hand of Fatḥ ʿAlī, King of his time".[14] This motto was reportedly composed by Čirāq ʿAlī Khān Navāʾi, the early 19th-century poet and statesman.[15]

It is one of several seals the Shah used in the course of his life. Three types of them were known to Jahāngīr Qāʾim-Maqāmī. The first is oval in shape and dates from the time when future Fatḥ ʿAlī Šāh was officially a crown-prince under the name of Bābā-Khān. It bears the legend "عبده الراجی باباخان ۱۲۱۱"—"His slave, who has trust in Him, Bābā Khān, 1211 (1796–1797)". The second was pear-shaped, bearing the legend "العزة لله + شاه شاهان جهان فتحعلی"—"The glory belongs to God! King of the kings of the world Fatḥ ʿAlī".[16] The third is the same as we have on our *farmāns*.

Karīmzāde Tabrīzī mentions two more seals. There is another one from the time before Fatḥ ʿAlī Šāh's enthronement, which has a rectangular shape and is inscribed "لا اله الا الله الملک لله عبده باباخان"—"There is no god but Allāh, the kingdom belongs to Allāh, His slave Bābā Khān". After thirty years of his reign had elapsed, Fatḥ ʿAlī accepted the title of "*ṣāḥib-qirān*" that was reflected in the new seal's legend which is similar in shape to the rectangular seal from our *farmāns*: "الملک لله سلطان صاحبقران⭑ گرفت خاتم شاهی ز قدرت ازلی ⭑ قرار در کف شاه زمانه فتحعلی ۱۲۴۴"—"The kingdom belongs to

14 J. Qāʾim-Maqāmī, "Muhrhā, tughrāhā va tawqīhā-yi pādshāhān-i Īrān az Ilkhānān tā pāyān-i Qājāriye", 1348 (1970–1971), 56, figs. 75, 76, 78; M. ʿA. Karīmzāde Tabrīzī, *Seals, tugras and firmans of Persian kings* [in Persian], London, 2006, 196.

15 Qāʾim-Maqāmī, "Muhrhā, tughrāhā va tawqīhā", 1348 (1970–1971), 32.

16 The ring with this seal is now kept in the Nasser David Khalili collection (M. ʿA. Karīmzāde Tabrīzī, *Seals, tugras and firmans*, 2006, 199).

God! Sultan, the Lord of Auspicious Conjunction. The royal signet ... (etc., the date is 1244–1829/30)".

On the reverse side of the *farmāns* there are seals and signatures of various authorities, their number varying from ten to just one seal and signature of Riḍa Qulī Khān Navā'i, *munšī al-mamālik*. The repertory of seals is similar to the examples given by Karīmzāde Tabrīzī.[17]

As it has already been mentioned, 23 of the 24 documents issued in the name of Fatḥ ʿAlī Šāh (with the exception of Pers. Doc.34), were written on blue Russian watermarked paper (Pers. Doc.14–33, 35–37). European watermarked paper first started to be used regularly in the chancelleries of the Safavid rulers. The analysis of over a thousand *farmāns* and *raqams* (edicts of princes) from the Qājār period, mostly preserved in the Institute for the Iranian Contemporary Historical Studies in Tehran, was undertaken by Riḍā Farāsatī, who showed that all of the 20 *farmāns* from Fatḥ ʿAlī Šāh's reign that were studied by him had been written on watermarked paper, and so are 30 of the 36 *raqams*.[18] Besides, there is a clear tendency towards the choice of colored paper,[19] which, probably, became the reason why Russsian blue or grayish-blue paper was in particular demand in the chancellery of Fatḥ ʿAlī Šāh.

The paper on which the set of documents is written belongs to the types that were quite common in Russia in the early 19th century and have been well described and catalogued. Before we continue with their identification, it should be mentioned that studies in the history of the paper production in Russia and cataloguing of watermarks found in Russian manuscripts and documents began in the 19th century, the works of the historian and palaeographer Nikolay Lihachôv (1862–1936) marking the most important advance in this field.[20] Fundamental research involving a wide range of archival sources was continued in the second half of the 20th century in the works by Zoâ

17 M. ʿA. Karīmzāde Tabrīzī, *Seals, tugras and firmans*, 2006, 213. The reading of legends of seals and signatures is given in Âstrebova, *Persidskie i tadžikskie dokumenty v Otdele rukopisej Rossiyskoj Natsional'noj biblioteki*, 1999, 28–54, 56–60.

18 R. Farāsatī, "Āšnāyī bā farmānhā va raqamhā-yi dawra-yi Qājāriye (taẕhīb va kāghaz)", 1379/2001, 242.

19 R. Farāsatī, "Āšnāyī bā farmānhā va raqamhā-yi dawra-yi Qājāriye (taẕhīb va kāghaz)", 1379/2001, 232 and 243.

20 N. P. Lihachôv, *Bumaga i drevnejshie bumažnye mel'nitsy v Moskovskom gosudarstve: Istoriko-arheografičeskij očerk*, 1891; id., *Paleografičeskoe značenie bumažnyh vodânyh znakov*, 1899, pt. 1–3.

Uchastkina (1906–1975)[21] and Sokrat Klepikov (1895–1978)[22], whose publications still remain the essential source of reference for contemporary scholars.

The table attached to this article [see Table] shows the presence of the three important elements of the Russian watermarks as seen in our *farmāns*.[23] The main, pictorial part is usually represented with a coat of arms of one of the Russian cities or provinces where the paper mill was situated, or of the mill's owner; it is placed in the right half of the sheet. The letters (usually placed in the left half) are the initials of the factory itself and its owner's (or owners') name. Finally, there is the year of production, placed in the middle of the lower part.

The tendency to mark the Russian goods in general with special signs proving their domestic provenance and indicating the mill where they were produced dates back to the second half of the 17th century. The first appearance of date on the Russian paper occurred in 1742 (Krasnoe Selo Mill). In 1744 a decree was issued that made obligatory the marking of all "goods of better quality" produced in Russia, the rules for such marking to be worked out by the Board of Manufacturers. Since not all the manufacturers of paper duly observed the rules, in 1778 a new decree was issued by Catherine II that pertained specifically to paper-making and had the title "On the manufacture of paper with an indication of the mill and the year of manufacture". It made obligatory the inclusion of the emblem of the mill, the name of the owner and the year of production into the watermark, and a massive effort was made to provide for the strict observance of the decree.[24]

5 Watermarks

5.1 *Watermark: Coat of Arms of Yaroslavl'.* [*Fig. 9.1, Pers. Doc.21*]

Twenty of the represented *farmāns* are written on the paper produced by the mills in and near the city of Yaroslavl'.

A number of sheets were made at the mill of the Yaroslavl' Large Manufactory (Bol'shaâ Âroslavskaâ Manufaktura). The textile factory in Yaroslavl' was established in 1722 under the order of Peter I by the Dutchman Ivan Tames and

21 Her best known monograph was published in Engish: Z. V. Uchastkina, *History of Russian hand paper-mills and their water-marks*, 1962.

22 See: S. A. Klepikov, *Filigrani i štempeli na bumage russkogo i inostrannogo proizvodstva XVIII–XX veka*, 1959; id., *Filigrani na bumage russkogo proizvodstva XVIII–nachala XX veka*, 1978.

23 See S. A. Klepikov, *Filigrani i štempeli*, 1959, 13.

24 Z. V. Uchastkina, *History of Russian hand paper-mills*, 1962, 163–165.

the local merchant Mihail Semônovich Zatrapeznov who soon became the sole owner. In 1731 he also started a paper mill. In 1764 the Manufactory was acquired by the famous businessman and self-made man Savva Âkovlevich Âkovlev (Sobakin) (1712–1784),[25] one of the richest merchants and fabricants of Catherine II's epoch.[26]

After the death of Savva Âkovlev his immense fortune worth 7 million rubles became the object of claims and litigation on the part of his numerous heirs, until finally the division was made between his widow, three sons and the children of the late fourth son, Mihail, by the special order of Catherine II.[27] The Yaroslavl' Large Manufactory passed into the possession of the children of the deceased Mihail. The following combination of letters found in the *farmāns* points at the provenance of paper from the Yaroslavl' Large Manufactory:

ЯМВСЯ (Pers. Doc.16, 22, 25, 27, 29–30, 32, paper dated 1802–1803)—"Âroslavskaâ Manufactura vnukov Savvy Âkovleva" (the Yaroslavl' plant of Savva Âkovlev's grandchildren);[28]

One of the documents (Pers. Doc.31, paper dated 1802) shows the combination not registered by S. A. Klepikov: **ЯМБ**—probably, "Âroslavskaâ Manufaktura Bumagi" (Yaroslavl' paper-mill).

A very similar design is found in a range of documents with the following combination of letters in the watermark:

ЯМГІЯ (Pers. Doc.14–15, 17–21, 24, 26, 28, 33, 35, paper dated 1801–1803)—"Âroslavskaâ Manufaktura Gospodina Ioanna Âkovleva" (the Yaroslavl' plant of Master Ioann Âkovlev).

The watermark refers to the mill founded on the Kotorosl' river in Velikoe Selo near Yaroslavl' around 1750 and bought by Savva Âkovlev in 1776. Ioann, Savva Âkovlev's grandson and the son of Mihail Âkovlev (d. before 1784), inherited it, and after the village was transferred from the jurisdiction of Rostov to the Yaroslavl' district in 1798, he substituted the earlier emblem of the Rostov coat of arms with that of Yaroslavl'.[29]

The pictorial part of the watermark in all the above-listed documents includes two variations of the coat of arms of the city of Yaroslavl' and its province. In the first variant the composition includes an oval with a bear standing

25 Z. V. Uchastkina, *History of Russian hand paper-mills*, 1962, 39–40.

26 See his biographical sketch: A. B. Muhin, "Savva Âkovlev—kupec, promyšlennik, predprinimatel'", 2005.

27 A. B. Muhin, "Savva Âkovlev", 2005, 170.

28 See Z. V. Uchastkina, *History of Russian hand paper-mills*, 1962, plates 15–17; S. A. Klepikov, *Filigrani na bumage*, 1978, 236.

29 Z. V. Uchastkina, *History of Russian hand paper-mills*, 1962, 77–78, plates 20 and 21.

FIGURE 9.1 *Watermark of Pers. Doc.21. Sketch Olga Yastrebova*

on its hind legs with a halberd in his left front paw, facing left; the oval is surrounded on both sides by branches tied with a ribbon at the place of their intersection; the imperial crown rests on top of the oval.[30] One of the documents has a depiction of the bear with a halberd with a crown on top, but without an oval or branches.[31] This sheet is also remarkable for not bearing the year of production. Paper of other documents from this group has dates 1801, 1802 and 1803.

Fourteen of the documents have additional "Arabic" or "Roman" digits (4, II, IV, V and VI) placed in the upper part of the sheet on the central axis. They obviously signify the grade of paper, the best grade marked by number 1 being absent among the *farmāns*.[32]

30 Compare types 8, 9, 12, 14, in: S. A. Klepikov, *Filigrani na bumage*, 1978, 237.

31 S. A. Klepikov, *Filigrani na bumage*, 1978, type 11 (but facing left).

32 Z. V. Uchastkina, *History of Russian hand paper-mills*, 1962, 51.

FIGURE 9.2 *Watermark of Pers. Doc.23*

5.2 *Watermark: Coat of Arms of Kostroma. [Fig. 9.2, Pers. Doc.23]*

The pictorial part of the watermark of two documents (Pers. Doc.23, 1804, and Pers. Doc.36, 1809) represents the emblem of the Kostroma province, a sailing ship (a galley, according to the official description of the arms) on water.[33]

The letters accompanying the arms are the following:

КГКОСN ДМ (Pers. Doc.23, paper date 1803)—"Kostromskoj gubernii Kinešemskoj okrugi selo Novodemidovskoe Demida Meŝaninova" (of Kostroma province, Kinešma district, Novodemidovskoe village, Demid Meŝaninov's [mill]);

КГКОСN ММ (Pers. Doc.36, paper date 1807)—"Kostromskoj gubernii Kinešemskoj okrugi selo Novodemidovskoe Markela Meŝaninova" (of Kostroma province, Kinešma district, Novodemidovskoe village, Markel Meŝaninov's [mill]).[34]

33 S. A. Klepikov, *Filigrani i štempeli*, 1959, 17, fig. 16 Г.

34 S. A. Klepikov, *Filigrani i štempeli*, 1959, 50, no. 281 and 182; id., *Filigrani na bumage*, 1978, 27, no. 340–344, 79, notes 211, 213.

The mill was built in Adiŝevo village in 1752 by the merchant Danila Zemskov. In 1778 it was acquired by Demid Meŝaninov (1726–after 1811), who renamed the village, after himself, to Novodemidovskoe.[35] The new owner was a scion of a rich and influential family of merchants from Kolomna and in 1783–1786 even held the post of the Moscow city mayor (*gorodskoj golova*).[36] Klepikov mentions the earliest watermark of Demid's son Markel as bearing the date of 1804, so somewhere around this time the mill must have passed to him.[37] The paper plant in Adiŝevo is still functioning.[38]

5.3 *Watermark: Coat of Arms of the Maslov Family.* [*Fig. 9.3, Pers. Doc.37*]

The last and the latest *farmān* (Pers. Doc.37) dated 1809 was written on paper bearing a watermark containing the date 1808 and the depiction of the coat of arms of the noble family of Maslov with the Biblical figure of Samson rending the lion's jaws on the blazon. The letters are ***сутф им***—"Sengileevskogo uezda Terenginskaya fabrika Ivana Maslova" (Ivan Maslov's plant in Teren'ga of the Sengilej *uezd*).[39] The *uezd* of Sengileĭ was part of the Simbirsk province in the Volga region, South of Nizhnij Novgorod. Teren'ga (see the map below) was founded in 1682 by the order of Tsar Feodor Alexeevič, and since the 1740s it was owned by the princely family of Golicyn, who built a cloth factory there. Early in the 19th century the estate passed to a distant relative of the Golicyns, Chamberlain Ivan Nikolajevich Maslov (d. 1814). Uchastkina mentions that the mill was owned by him in 1809 and produced about 21,020 reams of paper.[40]

6 Observations and Conclusions

The study of paper in Fatḥ ʿAlī Šāh's *farmāns* from the time of the Russo-Persian war of 1804–1813 in the collection of the National Library of Russia confirms

---

35 Z. V. Uchastkina, *History of Russian hand paper-mills*, 1962, 125.

36 A. I. Aksenov, *Genealogiâ moskovskogo kupečestva XVIII v.* (*Iz istorii formirovaniâ russkoj buržuazii*), 1980, 60, 109–110.

37 S. A. Klepikov, *Filigrani na bumage*, 1978, 79–80, note 213.

38 See the official website of the Administration of the Ostrovskij area of Kostroma region: http://www.ostrovskoe.ru/index.php/ekonomika/ekonomika-rajona/vypuskaemaya-produktsiya (accessed 6 December, 2017).

39 S. A. Klepikov, *Filigrani na bumage*, 1978, 47, no. 757, 82, note 447; id., *Filigrani i štempeli*, 1959, 63, no. 577, 221.

40 Z. V. Uchastkina, *History of Russian hand paper-mills*, 1962, 138.

FIGURE 9.3 *Watermark of Pers. Doc.37*

that Russian blue paper of lower grades of quality was used extensively by the chancellery of Fatḥ ʿAlī Šāh during the first decade of the 19th century. Although some *farmāns* bear the same date and deal with similar events, these documents still display many differences in terms of the elements included in the watermarks of the paper, consisting of heraldic images, letters, the year of manufacture, and additional signs. None of the sheets were made with the same mould, although some show many similarities. The period between the date of production mentioned in the watermarks and the dates on which the Persian documents were issued varies from one to four years.

Of the manufacturers represented by this material, the Velikoe Selo and Yaroslavl' Large Manufacture's production predominate. Produced by the plants belonging to the Âkovlev family, the paper with the emblem of the bear was obviously by far the most actively imported into Iran during the early 19th century. All three places of production (Yaroslavl', Kostroma and Teren'ga) were situated in the Volga river region, which probably is not a mere coincidence but was connected with the convenience of transportation along the navigable waterway through Astrakhan.

Bibliography

Akty, sobrannye kavkazskoj arheografičeskoj kommissiej. 1866–1904. Tiflis, vols. 1–12.

Aksenov, Aleksandr Ivanovich. 1988. *Genealogiâ moskovskogo kupečestva XVIII v. (Iz istorii formirovaniâ russkoj buržuazii)*, Moscow, Nauka.

Âstrebova, Ol'ga Mihajlovna. 1999. *Persidskie i tadžikskie dokumenty v Otdele rukopisej Rossiyskoj Natsional'noj biblioteki: Sistematičeskij katalog*, St. Petersburg, Rossijskaâ nacional'naâ biblioteka.

Atkin, Muriel. 1980. *Russia and Iran: 1780–1828*, Minneapolis, University of Minnesota Press.

"Bagrationi, Tejmuraz Georgievič". 2005. *Bol'šaâ Rossijyskaâ Enciklopediâ*, Moscow, Naučnoe izdatel'stvo "Bol'šaâ Rossijyskaâ Enciklopediâ", 637.

Bournoutian, George. "Alexander, Prince", *Encyclopaedia Iranica* 1/8, 826; an updated version is available online at http://www.iranicaonline.org/articles/alexander-prince-known-in-persian-as-eskandar-mirza-pro-persian-member-of-the-royal-family-of-georgia-b (accessed on December 2017).

Dvorânskie rody Rossijskoj imperii. 1996. Vol. 3: Knâz'â, ed. by Stanislav Dumin, Moscow, "Likominvest".

Farāsatī, Riḍā. 1379/2001. "Āšnāyī bā farmānhā va raqamhā-yi dawra-yi Qājāriye (taẕhīb va kāghaz)", *Tārikh-i mu'āṣir-i Īrān* 13–14, 233–244.

Farāsatī, Riḍā. 1390/2012. "Vārid kardan-i asbāb-i kārkhāne-yi kāghazsāzī az Rūsiye dar dawre-yi Nāṣirī (nāmehā-yi Mīrzā Āqākhān Nūrī va Mīrzā Sa'īd Khān-i Anṣārī be Maḥmūd Khān Nāṣir al-mulk)", *Nāme-yi Bahāristān* 12/18–19, 387–404.

Ismajyl, Togrul. 2006. *Istoričeskaâ geografiâ Azerbajdžana v XVII–XVIII vv.*, Baku.

Karīmzāde Tabrīzī, Muḥammad 'Alī. 2006. *Seals, tugras and firmans of Persian kings*, London.

Klepikov, Sokrat Aleksandrovič. 1959. *Filigrani i štempeli na bumage russkogo i inostrannogo proizvodstva XVIII–XX veka*, Moscow.

Klepikov, Sokrat Aleksandrovič. 1978. *Filigrani na bumage russkogo proizvodstva XVIII—nachala XX veka*, Moscow, Nauka.

Lang, David Marshall. 1957. *The last years of the Georgian monarchy, 1658–1832*, New York, Columbia University Press.

Lihachôv, Nikolaj Petrovič. 1891. *Bumaga i drevnejshie bumažnye mel'nitsy v Moskovskom gosudarstve: Istoriko-arheografičeskij očerk*, St. Petersburg.

Lihachôv, Nikolaj Petrovič. 1899. *Paleografičeskoe značenie bumažnyh vodânyh znakov*, St. Petersburg.

Muhin, Aleksandr Bronislavovič. 2005. "Savva Âkovlev—kupec, promyšlennik, predprinimatel'", *Vestnik SPbGU* Ser. 8/4, 153–171.

Qā'im-Maqāmī, Jahāngīr. 1348 (1970–1971). "Muhrhā, tughrāhā va tawqīhā-yi pādshāhān-i Īrān az Ilkhānān tā pāyān-i Qājāriyeh", *Barrasīhā-yi tārīkhī* 4/2–3, 123–162; 4/4, 15–67.

Uchastkina Zoâ Vasil'evna. 1962. *History of Russian hand paper-mills and their watermarks*, Hilversum, The Papers Publication Society, "Monumenta chartae papyricae historium illustrantia" IX.

TABLE *Description of watermarks of the Persian* farmāns *on Russian paper (collection 1236 "Persian Documents", NLR)*

Pers. Doc. no.	Date of the document	Heraldic image of the watermark	Letters	Year	Additional signs	Publication of the document in *Akty*
14	19 Rağab, 1218/ 5 November, 1803	Yaroslavl' Coat of Arms	ЯМГІЯ	1801	"4"	II, 803, № 1653
15	Ḏū al-ḥiğğa, 1218/ March–April, 1804	*idem*	ЯМГІЯ	1801	"4"	II, 804, № 1654
16	Rabīʿ II, 1219/ June–August, 1804	*idem*	ЯМВСЯ	1802		
17	Ğumādā I, 1219/ August–September, 1804	*idem*	ЯМ ГІЯ	1802	"4"	II, 815, № 1682
18	Ğumādā I, 1219/ August–September, 1804	*idem*	ЯМ ІЯ	1802	"4"	II, 817, № 1684
19	Ğumādā I, 1219/ August–September, 1804	*idem*	ЯМ ГІЯ	1801	"4"	II, 816, № 1683
20	Ğumādā I, 1219/August–September, 1804	*idem*	ЯМГІЯ	1801	"4"	II, 812, № 1673
21	Ğumādā I, 1219/ August–September, 1804	*idem*	ЯМГІЯ	1801	"4"	II, 818, № 1685
22	Ğumādā II, 1219/ September–October, 1804	Yaroslavl' Coat of Arms, without shield	ЯМВСЯ	–		II, 820, № 1689

TABLE (suite)

Pers. Doc. no.	Date of the document	Heraldic image of the watermark	Letters	Year	Additional signs	Publication of the document in *Akty*
23	Šaʿbān, 1219/ November–December, 1804	Kostroma Coat of Arms	КГКОСН ДМ	1803	Roman "II"	II, 818, № 1686
24	Šaʿbān, 1219/ November–December, 1804	Yaroslavl' Coat of Arms	ЯМГIЯ	1802	Roman "II"	II, 818, № 1686 (appendix)
25	Šaʿbān, 1219/ November–December, 1804	*idem*	ЯМВСЯ	1803		II, 818, № 1687
26	Šaʿbān, 1219/ November–December, 1804	*idem*	ЯМ ГIЯ	1803	Roman "VI"	II, 818, № 1687
27	Ramaḍān, 1219/ December, 1804–January, 1805	*idem*	ЯМВСЯ	1802		II, 819, № 1688
28	Ramaḍān, 1219/ December, 1804–January, 1805	*idem*	ЯМ ГIЯ	1802	Roman "VI"	II, 819–820, № 1688
29	Ramaḍān, 1219/ December, 1804–January, 1805	*idem*	ЯМВСЯ	1802		II, 819–820, № 1688
30	Ramaḍān, 1219/ December, 1804–January, 1805	*idem*	ЯМВСЯ	1802		II, 819–820, № 1688
31	Ramaḍān, 1219/ December, 1804–January, 1805	*idem*	ЯМБ	1802		II, 819–820, № 1688
32	Ramaḍān, 1219/ December, 1804–January, 1805	*idem*	ЯМВСЯ	1802	Roman "V" above	II, 819–820, № 1688
33	Ramaḍān, 1219/ December, 1804–January, 1805	*idem*	ЯМГIЯ	1803	Roman "VI" above	II, 819–820, № 1688

Pers. Doc. no.	Date of the document	Heraldic image of the watermark	Letters	Year	Additional signs	Publication of the document in *Akty*
34	Ramaḍān, 1219/ December, 1804–January, 1805					II, 639, № 1284
35	Šawwāl, 1219/ January–February 1805	*idem*	ЯМГІЯ	1801	"4" above	II, 812, № 1673
36	Rabīʿ al-āḫir, 1224/ May–June, 1809	Kostroma Coat of Arms	КГКОСН ММ	1807		III, 506, № 894
37	Ǧumādā I, 1224/ June–July 1809	The Maslov Coat of Arms	*сутф*	1808		III, 506, № 893

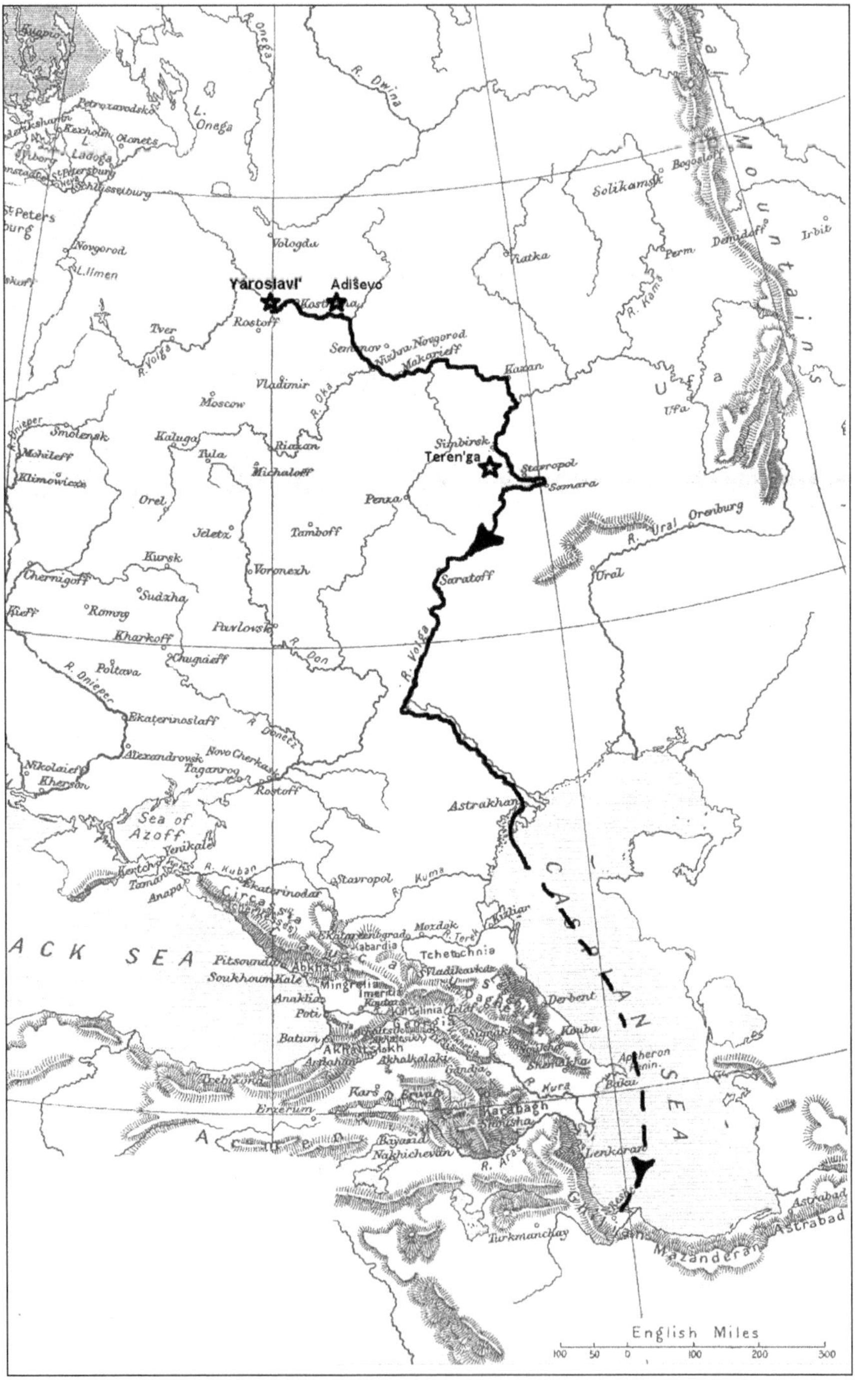

MAP 9.1 *Location of the Russian paper-mills discussed in the article and probable route of their products' delivery to Iran*

BY OLGA YASTREBOVA AND IGOR GUROV

Indices

Index 1: Countermarks, Dry Seals, and Type of Language and Script

Index 2: Watermarks, Dry Seals Motives

Index 3: Producers, Initials, Mills, Trade Companies, and Traders

Index 4: Kinds of Paper and Conditioning

Index 5: Toponyms, Empires, Communities, and Places of Production

Index 6: Manuscripts

Manuscripts on which the studies of the chapters in this volume are based are in Arabic language and script, apart from thc following:

Library, collection

Index 7: Personal Names (other than Producers of Paper)